On Appearance

ISSUE EDITORS: RICHARD GOUGH & ADRIAN KEAR

Forthcoming Issues

Performing Literatures

Vol.14 No.1

The relationship between text and performance has often been seen as pivotal to the notional division between theatre and live art. If theatre is drama is literature, performance is performer is liveness. In the academy, with UK theatre departments emerging historically from literature departments, their propulsion away from over-reliance on text has led logically enough to a preoccupation with the *non-text-based*. But are any of these distinctions as neat as is often assumed? *Non-text-based* performance is often heavily reliant on text; while *literary* drama is frequently more concerned with the live and processual than many would allow. *Performing Literatures* takes a fresh look at the academic and professional landscape, attempting to reconsider the text/performance dichotomy from a variety of critical perspectives.

On Training

Vol.14 No.2

This issue brings together reflections and analyses of training from a variety of specialisms and standpoints, and from various different international cultures and practices. Accounts of particular training regimes – dance, the gym, therapy – sit alongside polemical accounts of what happens to performers when they're trained – and what they're trained for. Overview essays suggest and describe generic characterisations of training, its structures and modes of operation. At the same time, specialist studies aim to define how training works, on both body and mind, in order to produce that very particular entity, the trained subject. Lastly, but not least, the issue explores, for both its subjects and indeed its watchers, training as a pleasure. (Photo by Veronique Chance, one of the authors in this issue)

Submissions

Performance Research welcomes responses to the ideas and issues it raises. Submissions and proposals do not have to relate to issue themes. We actively seek submission in any area of performance research, practice and scholarship from artists, scholars, curators and critics. As well as substantial essays, interview, reviews and documentation we welcome proposals using visual, graphic and photographic forms, including photo essays and original artwork which extend possibilities for the visual page. We are also interested in proposals for collaborations between artists and critics. *Performance Research* welcomes submissions in other languages and encourages work which challenges boundaries between disciplines and media. Further information on submissions and the work of the journal is available at: http://www.performance-research.net or by e-mail from: performance-research@aber.ac.uk.

All editorial enquiries should be directed to the journal administrator: Sandra Laureri, *Performance Research*, Centre for Performance Research, The Foundry – Penglais Campus, Aberystwyth, Ceredigion, SY23 3AJ, UK. Tel: +44(0)1970 628716; Fax: +44(0)1970 622132.

e-mail: performance-research@aber.ac.uk

www: http://performance-research.net

ISSN 1352-8165

Typeset at The Design Stage, Cardiff Bay, Wales, UK and printed in the UK on acid-free paper by Bell & Bain, Glasgow.

Abstracting & Indexing services: *Performance Research* is currently noted in *Arts and Humanities Citation Index*, *Current Contents/Arts & Humanities* and *ISI Alerting Services*.

Editorial
On Appearance

ADRIAN KEAR

According to philosopher Giorgio Agamben, etymologically the term 'appearance' belongs to the same group of terms as 'image' (*spectrum*), 'mirror' (*speculum*), 'spectacle' (*spectaculum*), 'sign' (*specimen*) and 'beautiful' (*speciosus*). Deriving from a root 'signifying "to look, to see"', appearance denotes a category of being whose essence lies in 'giving itself to be seen' (Agamben 2007: 56). In fact, for Agamben, appearance is the essence of *species* - that special category of being whose essence coincides with its appearance, its being given to be seen, whose visibility provides the very grounds of its intelligibility (2007: 57). This may come as no surprise to those used to defending appearances against the reductive, often mistaken, view that they serve to mask or hide 'reality' or 'truth' or to those actively engaged in processes of facilitating the communication of a specific moment of becoming - of being's becoming image - through the theatre or performance event. Yet the philosopher's language might prove useful in examining the specificities of theatre as the locus of appearance *par excellence*, Likewise, the dynamics of theatrical presentation might be usefully examined in terms of their interweaving of the special and the specious, the surprising and the vacuous, the appearances that matter and yet have no matter other than 'mere' appearance.

On Appearance sets out from the conviction that appearance matters, and matters as the very 'stuff' that provides the species 'theatre and performance' with its substance, specificity and specialness. Exploring the role appearance plays in a range of cultural forms - from body art to live TV, shamanic invocation to video installation, magic show to 'non-professional' performance - *On Appearance* charts the construction, circulation and contestation of some of the imagined possibilities, lived realities, political identifications and creative opportunities opened up by thinking through the logic of appearance. It examines the correlation between modes of appearance and practices of disappearance and investigates their inscription in the recuperative dynamics of power. *On Appearance* explicates the practical, philosophical and political implications of Agamben's invocation of 'the task of politics' as being 'to return appearance itself to appearance, to cause appearance itself to appear' (Agamben 2000: 95).

How, then, does appearance itself appear in this issue? In what ways, and in what forms, is appearance figured, thought and practised? Who and what cause appearance to appear? And who, and what, are disappeared in the process?

The contributors to *On Appearance* navigate these and related questions. Both Kear and Harris frame the dynamics and modalities of appearance as essentially political - appearance is 'the introduction of a visible in the field of experience which then modifies the regime of the visible' (Rancière 1999: 99). For Kear, writing about the relationship between political intervention and theatrical image-making, the appearance of politics occurs in the gap between presence and representation. Through examining the theatricality of photographic and video works by

Performance Research 13(4), pp.1-3 © Taylor & Francis Ltd 2008
DOI: 10.1080/13528160902875572

Phil Collins, works involving the apparent 'self-presentation' of real people, he demonstrates that in a theatrical context presentation is always cross-cut by representation and as such theatricality involves the bringing to appearance of their relation - a relation of non-relation - in the space and time of the image. For Harris, the appearance of 'real people' on stage provides the opportunity to interrogate the cultural-political dynamics of 'authenticity'. Concerned that physical and social appearance might produce a presumptive normalization designed to contain political possibility and limit the distribution of the sensible (Rancière 2004: 12), Harris argues for a disruption of performance's aesthetic regime and performance studies' classificatory systems, presenting a case for recognizing the importance of the equality of intelligences of performers, audience and makers in the theatrical event. In Quarantine's *Susan and Darren*, Harris sees evidence of the need to critically evaluate the gap between ontology and performance, appearance and theatricality. Theatre emerges as a place where appearances are constructed, spectatorship is activated and politics negotiated. This would also seem to be the case for aladin, whose personal narrative interweaves an account of post-colonial migration with the practice of prestidigitation, effecting through both a 'magical' resolution of impossible situations and insoluble identifications.

For Bayly, appearance is, in a theatrical context, literally the appearance of someone. Philosophically, Bayly's point of departure for considerations of faciality in epistemologies of performance is Levinas, for whom the face 'cannot appear, it just faces'. For Levinas, theatre is not a 'face-to-face' relation but rather a site of 'sensible appearance': a thoroughly mediated domain facilitating the recognition of 'the impossibility of a pure appearing or a simple revelation'. Bayly observes, accordingly, that 'an ethics of appearing' needs to be thought 'as a supplement to the politics of representation'. He elaborates how this might work through examining the logic of expression operative in Duchene's photography, which he sees as a primitive theatre, a site of the 'inscription of the theatrical'. The photograph is, for him, a theatrical 'set-up' par excellence, and Duchene's experiments a sort of 'experimental theatre' demonstrating the appearance of the face as an index of subjective expression and defacement as a mark of political de-subjectification. Drobnik similarly examines the ethical dynamics of spectatorship through the lens of the photographic image, suggesting that ethical involvement in spectatorship is evident in theatrical events that create an association between the violence of an event - its incarnation 'outside' a theatrical context - and the violence of its appropriation within a frame of 'authored' representation. He notes that although spectators cannot intervene in the representation, they also cannot just consume; the latency of the event, of the documentary face-to-face relation, is troubled by, and troubles, the dynamics of theatrical distanciation to create an economy of aesthetic implication in, and indifference to, 'real' suffering. Kelleher, similarly, explores how theatricality resides in the gesture of distancing the proximity of the performance taking place in front of the spectator. Focusing on the phenomenology of La Ribot's performance of laughter, Kelleher demonstrates how the interplay of duration and repetition forces recognition of a relation being forged within the work of theatre; not one of correspondence - 'their world is not our world' as spectators after all - but of fractured allusion; theatrical violence and political violence are not contiguous, even if they use the same discourse; theatrical suffering is not the same as suffering as such. Theatre thereby re-emerges as a world of appearances - 'and human appearances at that' - created with an intensity that both exceeds and undermines intentionality. Durkee likewise examines dance as the appearance of appearance rather than the appearance of a particular subject. Exploring the challenges to spectatorship attendant on apprehending a body-turned-object without reclaiming it as authentic, she locates a struggle between materiality and appearance operating in performance. She argues that although materiality cannot take possession of appearance, there is, nonetheless, an apprehension of the materiality of appearance manifest in the dialectical image.

For Kubiak, appearance is decidedly anti-materialist and must be reclaimed from the philosophical lexicon of materialism. Following Artaud and Blake, Kubiak insists upon theatre's visionary potentiality as a space for the revelation of 'possibilities not yet apprehended' and the 'visionary making and unmaking of a world'. For Kubiak, appearance is the appearance of a vision - a vision of 'redemptive, *political* life'. He invokes shamanic consciousness as means of re-envisioning the world with assistance from the 'apothecaries of the mind' and theatrical shamanism as a form of cleansing (catharsis) akin to taking the transformative natural hallucinogen ayahuasca. Lavery, likewise, allows imagination to drive his approach to making 'sensory appearance' appear at the level of embodied textuality, while for Tresize, the re-appearance of dead relatives in the live performances of John Edward, media medium, is evidence enough of the theatrical laundering of some of the oldest conjuring tricks in the book. She examines the work of 'making the disappeared reappear', demonstrating how the performances work by mechanizing appearance, but also revealing the mechanisms by which this is made possible in a postmodern construct linking the discourse of trauma to the technologies of the televisual. Tresize outlines an apparatus of appearance underpinned by an ideological consensus that 'erases the present' to create 'the appearance of appearance'. McGillivray traces this process back to the eighteenth-century construction of the picturesque landscape and its dependence on theatrical staging to produce aesthetic effect. The 'framing of the world' produced within this quintessential manipulation of scenographic compositional processes serves to ensure the world 'appears to be' as it is presented within the logic of the theatrical *mise-en-scène*.

For Gough, the ideological consensus governing the appearance of the figure of 'woman' in the political sphere is her rendering as 'spectre', unable to appear materially - politically - despite having massive political impact and import. In political liberation movements, 'Women still appear to disappear, or appear to be marginalized and interrupted'. Gough notes how, in Irish nationalist politics, 'when the call to Cathleen is heralded, 'real' women seem to vanish'. Further, she locates the appearance of actual activist women as the exception that allows for the continued exclusion of the category of 'woman' from practical politics. Gough accentuates the appearance and disappearance of available forms of subjectivity. Black female subjectivity, represented in the nineteenth century in the form of Aunt Jemima, 'is disappeared under the weight of the commodity's ubiquitous reproductions', while at the same time establishing the conditions of possibility for its twentieth-century political re-appearance. Roach similarly analyses the appearance of a different kind of heritage - 'intangible heritage' - and with it the appearance of a different kind of labour. Conceiving of sweat as the appearance of work, Roach notes that the history of slavery and exploitation also contains within it a heritage of labour - the extraordinary work (and works) of carnival and Mardi Gras. As a living tradition passed through family lines, the labour of making performance, and the material sweat of history this tradition contains, 'reappears in the faces of the next in line'. He argues that the work that emerges from conditions of hard labour - in free time, as 'free play' - allows alternative possibilities to appear and offers a glimpse of 'an exhilarating kind of freedom'.

REFERENCES

Agamben, Giorgio (2000) *Means without End*, trans. V. Binetti and C. Casarino, Minneapolis: University of Minnesota Press.

Agamben, Giorgio (2007) *Profanations*, trans. J. Fort, New York: Zone Books.

Rancière, Jacques (1999) *Disagreement*, trans. J, Rose, Minneapolis: University of Minnesota Press.

Susan and Darren
The appearance of authenticity

GERALDINE HARRIS

The publicity for *Susan and Darren* described it as an 'event with dancing' by Quarantine and Company Fierce (Quarantine 2008). *Susan* is Susan Pritchard. She is 50 years old, works as a professional cleaner and lives in Manchester sharing her house with her son Darren, who is a professional dancer and choreographer and the director of Company Fierce. These are not 'roles' that Susan and Darren play, it is who they are. Obviously, Darren has extensive on stage experience, but for Susan, as she confirms within the piece, this is both a 'first' and a 'one off'.

Originally, I intended to discuss *Susan and Darren* in relation to issues of ethnicity, sexuality, gender, age and the intersection of these categories with that of class. But when I started writing, this seemed rude, as in impolite, to Susan and Darren. This threw me because analysing shows in such terms is more or less what I do for a living. This essay therefore constitutes the unravelling of this response with reference to ideas from Jacques Rancière, but the process has also included discussions with Richard Gregory and Renny O'Shea, the artistic directors of Quarantine. While Rancière features heavily in the main text, I have included some of Richard's and Renny's remarks in the footnotes in a different `typeface`, where they function as interventions that correct my errors, provide 'off-stage' information about the piece and/or alternative, even contradictory perspectives. As such, rather than signalling an 'authoritative' account of *Susan and Darren* (not least because I have not spoken to Susan or Darren or to other members of the company), these inclusions are intended to underline the inevitable partiality of my reading.

I saw *Susan and Darren* at the Nuffield Theatre, Lancaster University in December 2006, a scheduling that effectively rendered it a 'Christmas Show' although the piece was not created for this purpose.[1] Moreover, the Nuffield doesn't really 'do' Christmas Shows because its programme is focused around small-scale touring productions of various genres but which collectively could be identified as 'experimental'. Using half the Nuffield space, the staging for the piece was roughly traverse with three rows of seats running down each side of the performing area. At one end the area was enclosed by a table about 10ft long by 5ft high, with nine or ten stools on one side and at the other by a half a dozen armchairs facing a row of video monitors. Otherwise, the set consisted of two banks of mobile disco lights, a chair and a 12ft, vertical dance pole. As the audience entered, they were encouraged to occupy the stools and armchairs.[2]

The event starts with Susan and Darren sharing autobiographical details and stories.[3] They seem open and candid, and we are invited into the centre of their daily lives, sometimes without introduction. For instance, they refer to 'Donna' and 'Keisha' in passing, assuming that, without being told, we will pick up that they are Susan's daughter and granddaughter and Darren's half-sister and niece respectively. Similarly, Darren asks Susan about his father, who we only later learn was killed in a violent

[1] There were big shifts in the show in response to context or particular changed circumstances, both at Susan's and Darren's home and in the performance space. For example, when we performed our 'Christmas special' in Lancaster, Darren was worried that he wouldn't get the 'Living Room' section right because Sue had moved everything around the weekend before in order to put their Christmas decorations up. He wouldn't countenance not being (or at least trying to be) faithful to what the reality of furniture layout was at home.

[2] Some of this description applies only to how we had to/chose to configure it at the Nuffield. Also you missed out the parquet dance floor, the four speakers, the CD player, the mic and stand, the mirror ball; they're our set too.

[3] It 'started' earlier than this. We (and Susan and Darren) always saw the dance workshop one hour before the show, the show and the party afterwards as inseparable parts of the whole 'event'.

Performance Research 13(4), pp.4-15 © Taylor & Francis Ltd 2008
DOI: 10.1080/13528160902875580

incident several months before he was born. Donna's father also died before her birth. Both of these men were Black, Susan is white.

In some sections of the show their dialogue appears fixed, at others, more spontaneous or rather 'rehearsed' only in so far as it touches on family stories, incidents and questions that have been told, asked and answered many times before.[4] Distinctly unrehearsed contributions are made by Susan's and Darren's friends and family through video recordings shown on the monitors, which sometimes run concurrently with on stage action. We hear from Darren's ex-boyfriend who still misses him and of Susan's bad times, her problems with men, with depression, her suicide attempts. This material is offset by lighter stories and by Susan and Darren dancing to disco classics,[5] and on one occasion they are joined by a dozen or so audience members who have been taught a routine just before the show. The pair also perform more formal and stylized sequences of movement, once together and three times Darren on his own.[6] Darren occasionally wanders into the rows of seats at the sides. Susan sits at the high table and enlists the aid of audience members to prepare a buffet: sandwiches, crisps, peanuts, squares of cheese on cocktail sticks. Two-thirds of the way through they declare a five-minute interval during which the audience can ask them any question they like. This occurs just before what Darren calls the 'serious bit', during which they each have a monologue. Darren's is about his father's death in a drunken brawl over a taxi on a piece of waste land on the edge of Moss Side, Manchester, and about Susan being raped on exactly the same piece of land four years later while *she* was waiting for a taxi. Susan's is a childhood story about being given a beautiful pink dress as a surprise Christmas present to wear to a school party. At the end of the show the audience are invited to buy a drink from the Nuffield bar, to share the buffet and to dance.

My attack of politeness in relation to *Susan and Darren* was the product of various factors. These include the fact that, as indicated, Susan and Darren are not playing roles but are performing (as) themselves. Yet this is virtually a norm for shows at the Nuffield, and previously I have, without qualm, analysed the work of artists who re-present 'themselves' on stage in relation to various theories of theatre and performance, subjectivity and identity. Notably however, there was such a feeling of 'authenticity' about this piece that afterwards one of my colleagues in Theatre Studies complained that Darren's monologue in the serious bit 'jarred' because it had been rendered into semi-poetic language.[7] Much of this sense of immediacy arises from the naturalness of Susan's and Darren's interaction with each other and with the audience but more specifically from Susan's 'appearance' on two counts. Her bodily style is that of an 'ordinary' middle-aged woman of the sort only rarely visible in professional theatre, performance or live art. Further, her stage presence shows up how far in these latter fields the presentations or performances of self,

Photo Simon Banham

4 The dialogue works in three main ways:
1) Fixed, scripted (e.g., Wasteland; Pink Bell Dress). It changes as much (or as little) as any text-based performance would.
2) Improvised within rigid, fixed frames (e.g., Living Room).
3) Rule-based, to maintain spontaneity, bring in new material, air subjects they wanted to confront in the context of performance (e.g., Darren's ongoing question to Sue on this tour about why she hasn't put him on the tenancy yet).

5 There is a fragment of disco music but the majority of music in the show is reggae and specifically Lover's Rock. This is music that Susan has listened to since her teens and was introduced to in the shebeens and blues clubs Darren talks about in the 'Wasteland' section.

6 Dancing is a huge part of Susan's and Darren's lives. Of the approximately 85-minute running time, 32 minutes were dance- or movement-based.

7 Sonia Hughes, the writer, argued for changing this section on the grounds that she was too visible in it. We argued for keeping it, for that very reason. Like some of the choreography and other sections of the show, the momentary movement away from the everyday was deliberate and important.

which in the mid-twentieth century were part of a rejection of 'theatrical technique', have become theatricalized even while they are still read as confusing the boundaries between the 'real' and the 'mimetic'. By contrast to most such re-presentations, there is no trace of theatrical self-reflexivity in Susan's demeanour. Instead, her self-consciousness is that of one with a role in a public ceremony that demands to be carried out with attention, seriousness and even, at times, solemnity.

It is the impression this gives of an encounter with Susan as a 'real person' that made it seem rude, even unethical, to take up the sort of distance from her that would be implied by analysing her (performance?) in terms of the politics of identity and above all in relation to class. This, not least, that in her book *Formations of Class and Gender* feminist sociologist Beverly Skeggs undertook an ethnographic study of a group of women living in the North West of England (where Susan, Darren, Richard, Renny and I all live), all of whom in classic Marxist-socialist terms might be deemed 'working class' but most of whom strongly refused this categorization (Skeggs 1997: 74-97). It is nevertheless Skeggs's project to re-open the question of class, which in her next book *Class, Self, Culture* she achieves for a 'post-Marxist' era in reformulated terms, which owe much to Foucault, Bourdieu and other 'post' theorists working within feminism, anti-racism and queer studies. She seeks to explore how certain values and meanings come to be 'fixed' to particular bodies or, as she puts it, 'how some forms of culture are condensed and inscribed onto social groups and bodies that mark them and restrict their movement in social space' (Skeggs 2004: 2). As part of this argument she identifies the way that bodies are read through classification systems in which class is not 'made alone' but alongside other classifications such as 'race' and gender (3). She stresses that such readings are based on the 'visible body' or, rather, on 'appearance' (156).

The problem is that I *do* think there are issues of class and other classifications systems in circulation around *Susan and Darren.* At the same time I have to admit the possibility of my reading these things onto it, at least partly on the basis of Susan's and Darren's 'appearance'. In short, not only ignoring the fact that they are performing but potentially re-naturalizing the very systems of visibility and categorization that I, like Skeggs, as a feminist wish to reveal as discursive fictions. This necessity of working within and therefore possibly re-naturalizing the socially constructed, hierarchical distinctions that one would critique, has been an acknowledged double bind of progressive politics informed by deconstruction. As a problematic, in the latter part of the twentieth century it fuelled an emphasis on the 'ethics of otherness', as a means of exploring the possibilities for a non-essentialist politics that could acknowledge and respect 'difference' while allowing for community and democratic consensus. Equally, it fuels the recent enthusiasm for the works of Alain Badiou and Jacques Rancière, both of whom have argued that in practice all our contemporary versions of ethics and the politics of difference ultimately come down to an identification, of and with the 'other' as a traumatized victim (see Badiou 2001: 11-14, Rancière 1999: 125-6). Considering the subject matter of the 'serious bit', this is another potential danger of reading of *Susan and Darren* in relation to the politics of identity.

In contrast to Skeggs, Badiou and Rancière are not concerned with revealing and analysing the power relations 'behind' the operations of what Rancière designates as the 'police', that is, 'the partition or distribution of the sensible' that governs the policing or ordering of places and occupations and determines what at any one time is 'commonly, doable, sayable, audible and visible' (see Rancière 1999: 29 and Rancière 2004a: 12-46). This because to do so risks politics being defined purely in the *terms* of the 'police', thereby foreclosing in advance the possibilities for resistance and political agency. Instead, Badiou and Rancière both focus on identifying

Photo Gavin Parry

the conditions under which 'new' political subjectivities emerge, which for Rancière are (at least temporarily) not already intelligible within the existing system of classifications, differentiation and inscription and which, therefore, cause a redistribution or reconfiguration of 'the sensible'.

In both instances, the generic nature of these 'conditions' have led to Badiou and Rancière being referred to as 'new universalists'. However, Rancière's argument is historical, specific to the context of Western democracy and identifies only 'one universal of politics', which is the presupposition of equality 'of one speaking being with any other speaking being' (Rancière 1999: 30). Even then, this 'equality' is neither a unifying principle nor a 'given' with an *a priori* meaning(s) but rather as a presupposition is something that must be verified continually (Rancière 1999: 31-5). For Rancière this is achieved through a process of 'dissensus'. Working on the understanding that the primary aim of democracy is to represent the entire *demos*, this refers to a dual process based on the notion that it is impossible for the police order to 'count' or identify all parts of the *demos* or population at any given time (see Rancière 1999: 22-3). A moment of dissensus describes a dis-identification with what is counted or recognized as subjectivity within the prevailing police order, on the basis of a 'wrong' and an 'impossible identification' of and with 'the part which has no part' in this count; impossible (partly) because on this basis it refers to a subject position that has been/is unintelligible (see Rancière 1999: 138-40).[8] However, since once it

[8] This is my own simplified interpretation of the 'impossible' nature of this identification (influenced by Judith Butler), which I am employing to avoid a lengthy and complex diversion from my current argument. Actually, in *Disagreement* and elsewhere, Rancière discusses this identification in terms of being attached to an 'empty signifier' by which he, like Ernesto Laclau, appears to mean a sort of rhetorical catachresis (see Laclau [1996], whose Chapter 3 is titled 'Why Do Empty Signifiers Matter to Politics?', pp. 36-44).

has been named 'this part' can be incorporated (literally, bodily) into the police order, it is the immediate and contingent *staging* or 'acting out' of a process of verification of equality that causes disruption and reconfiguration of the sensible (see Hallward 2006: 111).

For those of us struggling with the above double bind, there is much that is attractive as well as provocative in these ideas. They offer a means of side-stepping what Eve Kosofsky Sedgwick describes as the 'paranoid' register within certain feminist and queer theories which, like Rancière's own discourse, have been produced under the influence of Foucault but which ultimately only allow you to confirm what you already know (Sedgwick 2003: chapter 4). For some of us an added bonus with Rancière is that, as Peter Hallward indicates, theatre and performance have been core enabling concepts in his thinking (Hallward 2006). However, Hallward poses some 'strategic questions' about Rancière's thinking and suggests that Rancière sometimes overestimates his distance from 'the postmodern posture that he appears to oppose' (Hallward 2006: 125). Indeed, there are times when his ideas cover familiar territory and he shares far more ground with some feminist, post-colonialist and queer theorists than is acknowledged. This includes Judith Butler, and, as in her case, there is danger that Rancière's work could be recuperated to provide a springboard for over-enthusiastic claims for the 'inherent' politics of particular aesthetic strategies and/or for performance as a, if not the, privileged site of the political. In fact, like Butler, while deeply concerned with the relationship between politics and aesthetics, Rancière works to avoid the slippages within postmodern thought that effect either a simple reversal of the traditional hierarchy between the two or a collapse into indeterminacy where potentially all performance can be read as equally 'political' in the same way (see Rancière 1999: 32). As such, Rancière states in no uncertain terms that 'there is no criterion for establishing an appropriate correlation between the politics of aesthetics and the aesthetics of politics' (2004a: 62). Therefore, in drawing on Rancière to enable my discussion of *Susan and Darren*, I must stress that I am focusing primarily on his ideas on the *politics of aesthetics* in the arts, which are tied up with but not identical to his concept of the '*aesthetics of politics*' (see Rancière 2004a: 61-6).

Most of the work that tours to the Nuffield can be read in terms of the characteristics of Rancière's 'aesthetic regime of art', which he dates back to around the start of the nineteenth century. Essentially, a 'regime' is distribution of the sensible, a way of thinking about art that is implicated within the broader 'policing' of the demos (2004a: 45). The aesthetic regime represents a revolt against and reconfiguration of earlier regimes, which Rancière terms the 'ethical (or poetic) regime' and the 'representative regime', as established via the writings of Plato and Aristotle respectively (Rancière 2004a: 12-19, 25-30). Rancière proposes the concept of 'the aesthetic regime' as a means of displacing and reconfiguring the classification system represented by the terms 'modernism' and 'postmodernism', so as to create new perspectives on the debates they signify. As Hallward explains, the aesthetic regime refers to 'a fragile liminal state balanced between tendencies either to collapse the differences between art and non-art or to reify the gap between art and life' (2006: 121). This is not any sort of a guarantee of the *politics* of art, since as ever, for Rancière, 'politics' starts in a moment of dis-identification and unintelligibility that cannot be 'read' in terms of either pole and that ends when there is an identification with either extreme (Hallward 2006: 121). I am arguing that in small ways *Susan and Darren* disturbs what has become a common ordering of places and occupations within the field of performance under this aesthetic regime. As part of this, I am suggesting that in the particular theatrical context in which I saw the show, Susan's 'appearance' is, in effect, as one of 'the parts who have no part', at least in terms of the equality 'of one speaking being with any other speaking being' (Rancière 1999: 30, 138-40). While this is deeply embroiled with matters of class, gender, age and 'race', the disruption of the 'sensible' actually arises from a presupposition of equality in Quarantine's work between art and non-art but more importantly between artist and non-artist. It is this presupposition that accounts for my sense that it would be inappropriate to define Susan and Darren in terms of these normative identity classifications.

The participation of 'non-professional' performers in productions led by professional artists is, of course, nothing new.[9] Discussing *EatEat*, a Quarantine production made in Leicester in 2003 with a group of refugees and asylum seekers, Sally Doughty and Mick Mangan remark, 'From a distance the project looks like what we might lazily call "community theatre"' (Doughty and Mangan 2004: 32). By this they mean 'a theatre of social action', often more concerned with process than product, and they situate the genre through reference to Anne Jellicoe's large-scale, mid-twentieth-century community plays, created for and performed largely by geographically defined groups (32). However, Doughty and Mangan go on to say that *EatEat* is actually defined *against* this tradition, in so far as emphasis is placed the director's artistic vision and 'on production values framed within an aesthetic of contemporary western performance' (33).

Jellicoe's work is often discussed as part of a British post-1968 'alternative theatre movement'. As defined by Baz Kershaw in *The Politics of Performance*, this embraced a broad church of genres ranging from 'Carnival', in which Kershaw includes performance art, to community plays and agit-prop with much mixing and matching in between (Kershaw 1992: 82-4). However, commonalities often included the rejection of traditional theatre spaces and conventions, a drawing on 'popular' or 'folk' forms, or on social rituals, non-professional performers, audience participation and an emphasis on process over artistic product. Above all, according to Kershaw what characterized this 'movement' was that it was 'a theatre committed to bringing about actual change in specific communities' (1992: 5). This change might constitute political or social consciousness-raising, or be aimed at the production of a sense of what Victor Turner terms 'communitas', a 'direct, immediate and total confrontation of human identities' (Kershaw 1992: 28), or again, as Kershaw puts it in relation to Jellicoe, 'The celebration of unity - of a common identity overriding internal differences' (190).

At first glance, *Susan and Darren* does seem to belong to this tradition, via Susan's 'appearance', in drawing on popular forms and social rituals and the audience participation. However, the company's work tours to and is commissioned by theatre and performance venues such as the Nuffield, Contact in Manchester and the Tramway in Glasgow, unlike much community theatre *all* Quarantine performers are paid (where legal, see note 9) and there is a clear emphasis on 'product'. *Susan and Darren*'s, programme credited two directors, a designer, a choreographer, a writer, a lighting designer, a video artist, a sound designer and two producers.[10] There are some breathtaking theatrical moments in the piece, which include what, for want of a better term, I'll have to call Darren's 'pole dance',[11] a designation that belies its formal beauty and affect seen live, and a sequence in which Susan methodically washes his body section by section as he lies on the high table. Both of these segments depend partly on Darren's extraordinary appearance as a professional dancer but equally on the creative deployment of a wide range of the techniques and technologies that commonly underpin much experimental contemporary 'composition' in both dance and theatre. These sections also heavily lent themselves to being read 'intertextually' in relation to classical religious iconography.

As Doughty and Magnan conclude then, Quarantine's productions are as concerned with aesthetic experimentation and theatrical effect as those of many other companies whose work is

9 There is a problem of definition. 'Non-professional' or even 'amateur' often imply either unpaid or inept. Susan and Darren were neither. We pay all our performers (when we're allowed to: with EatEat's performers, this was illegal). 'Non-performer' is absurd, because they clearly are performers in the context they're encountered in (and if the argument is made that this isn't what they do most of the time, let me line up some thousands of self-defined 'actors' who haven't done any work paid or unpaid for donkey's years): 'untrained' is not specific enough, and what kind of training counts: RADA? BTEC? degree in theatre history? The workshop-hunting autodidact? We prefer Rimini Protokoll's term (or thereabouts) 'experts in everyday life'. Susan is there because nobody else could replace her.

10 One of the producers is Darren's sister Donna.

11 That's what we all call it too.

usually understood as engaged far more with formal concerns than with social or political intervention or 'communitas'. This would make my colleague's comment on the poeticization of Darren's speech in the 'serious bit' seem naïve, except that Quarantine's website offers press quotes describing their work as 'artfully directed *and* utterly authentic' and as 'The Wooster Group crossed with *The Osbournes*' and 'reality theatre' (Quarantine 2008, my emphasis).

Significantly, a while before seeing *Susan and Darren* I heard Richard Gregory give a talk about Quarantine during which he was asked about the ethics of artist-led productions based on the 'real life stories' of non-professional performers, which sometimes include painful material. Later in private conversation he confirmed that 'we are often asked that question'.[12] This did not surprise me because, having only heard about their shows and seen still pictures, from a twenty-first century perspective rather than community theatre the association I made was with 'reality TV'. Having seen *Susan and Darren*, this association still holds, on two seemingly contradictory counts. It holds because outside of some types of 'community art practice' - and unlike the vast majority of theatre and performance - television and especially reality TV constitutes a space where 'ordinary people' like Susan or some of the participants of *EatEat*, *White Trash* (2004) or *Butterfly* (2004) make appearances in the public realm (apparently) *speaking for and about themselves*. In this I am making another 'generic' distinction between Quarantine's work (and productions) where professional performers represent such groups, however 'verbatim', and the type of show I have seen on the experimental touring circuit, where 'non-professional' (usually unpaid) participants have appeared, mostly on video and only rarely live. In these instances, once again either artists have mostly spoken for or about these individuals or their inclusion is primarily employed as a vehicle for speaking about the *artist*. Yet equally, the comparison between Quarantine and 'reality TV' also holds in that, to a greater extent than with these types of 'verbatim' and experimental work, I have heard expressions of uneasiness over Susan's appearance in the show, which indicate anxieties about Quarantine's work in terms of 'behind the scenes' manipulation of its subjects and about voyeurism.

Actually, just like the alternative theatre movement, reality TV covers a broad range of practices, which can be described in terms of gradations and mixings between the extremes of 'Carnival' and 'agit-prop', or rather between 'pure' entertainment and 'serious' socially or politically committed documentaries and drama-documentaries. Further, commentators such as Ib Bondjeberg have argued that the massive expansion of this genre from around the late 1980s can be seen as a democratization of television as a conservative, patriarchal institution in terms of giving space and visibility to a greater diversity of the population than was previously the case (Bondjeberg 1996). However, while there are examples of questionable practice in this field, there is often a generalized suspicion of reality TV within academia as elsewhere, which reflects a larger suspicion of both the politics and aesthetics of television as a medium. This is most clearly evinced by the fact

[12] We have also asked this question of ourselves since we began working as Quarantine and continue to debate it.

Photo Simon Banham

that television is seldom accorded the status of 'art' *strictly in its own terms*. Despite the championing of its 'popular pleasures' and its potential to create resistant reading by figures like John Fiske (1999), television is still frequently defined as encouraging passive consumption of its images by spectators who are figured either as being deluded into accepting them for transparent reflections of reality, or as being wholly seduced by the 'hyper-reality' they construct. By extension, this medium has been held accountable for the aestheticization and evacuation of the political and the loss of the social and communal. Interestingly, those who articulate these views are not fooled or seduced by television themselves but rather express concern on behalf of 'ordinary viewers' or the 'masses', who are constructed therefore as far more gullible and less sophisticated than the critic in their ability to interpret the play of signs.

In terms of its particular mix of aesthetics then, within the theatre, Quarantine's work could be said to disturb some of the hierarchies and boundaries that Aristotle articulated in *The Poetics* and Rancière identifies as constituting the 'representative regime of art' (see Rancière 2004a: 21-2, 36-7). As Rancière indicates, Aristotle lays down 'rules' for the genres, mediums and forms 'proper' to certain types of subjects (in the double sense of stories and persons), on the basis of ethnicity, 'class' (position and reputation) occupation and gender (Aristotle 1975: 48, 51). These orderings were supposedly crossed and confused by both modernism and postmodernism. However, if they are effective at reconstituting themselves (and obviously, I am arguing that this is the case), this may due to the 'sets of relations resting on some key equivalences and some key oppositions' that Rancière identifies in 'The Emancipated Spectator' as having shaped the 'reform of the theatre' going back at least to Brecht and Artaud and which he asserts still do, 'even in postmodern guise' (Rancière 2004b: 4, 2). He identifies these as 'equivalence of theatre and community, of seeing and passivity, of externality and separation, mediation and simulacrum; oppositions between collective and individual, image and living, activity and passivity, self-possession and alienation' (4). According to Rancière, subject to 'rearrangement', the same terms underpinned Plato's condemnation of theatre and his preference for 'choreographic performance' (2-3). Above all, Plato distrusted theatre due to the way mimesis 'split' and doubled identity or 'self-possession' (Rancière 2004a: 13).

In developing his theme, Rancière points out that much modern (and postmodern) thinking around theatre has operated on the assumption that 'spectatorship', one of the qualifications *for* theatre, is a' bad thing'. This because it is characterized as 'passive' and because 'looking' is also 'a bad thing'; 'the opposite of knowing and of action, it means being in front of an appearance without knowing the conditions of production and without any power of intervention' (Rancière 2004b: 2). In reforming the theatre and attempting to overthrow the Aristotelian representative regime, efforts were therefore made to render the spectator active 'either by maximizing the distance between spectacle and spectator' (Brecht) or by minimizing it (Artaud)' (Hallward 2006: 114). Implicit to these positions is a presupposition of the possibility of direct transmission between the 'mastery' or 'knowledge' of the (active) artist and the 'ignorant' (passive) spectator. What actually emerges as the chief obstruction to these objectives is the *spectacle* or representation itself. It is the spectacle that produces the spectator as a spectator, one who passively looks at the surface of things, mere appearances, rather than one who acts or learns, and the spectacle is also that which 'stands between' the artist and the spectators. The desire is then to overcome this mediation, either by foregrounding its status as deceptive spectacle and revealing the conditions of is own production (Brecht) or transcending it so that it becomes 'life itself' (Artaud). In both instances, Rancière suggests that the presupposition against mediation is connected to 'the presupposition that the essence of theatre is the essence of the community ... by reason that uniquely on the stage, real living bodies give the performance for people who are physically present together in the same space' (2004b: 9). In ultimately a very Platonic fashion then, both these approaches aspire to suppress the 'bad' bits of theatre (which are the bits that *make* it 'theatre') so as to 'restore' the spectators to the state of being an active part of a living community, either through leading them 'to the position of a citizen who acts as a member of the collective' (Brecht) or to being 'carried away in a flood of collective energy' (Artaud) (Rancière 2004b: 9).

Rancière's exposition maps exactly onto Kershaw's description of the aims of the British alternative theatre movement, which do suggest a presupposition that the 'essence of theatre' is, could or should be, the essence of 'community', whether this essence is understood in social, political, anthropological or metaphysical terms. It also maps both onto the discourse around television as a medium discussed above and onto a conception of a postmodern theatre, self-reflexively endeavouring to reveal what is 'behind' its own appearance(s), whether to point to the unrepresentability of the real or to the discursive nature of all experience and identities. In fact, it might map onto any contemporary performance that has described itself or been described as aiming to 'activate' the spectator in some way, whether as part of a Lyotardian theatre of the sublime, an ethical theatre of witnessing or a political theatre of identity.

This is because, Rancière observes, the equivalences and oppositions listed above represent

> a partition of the sensible, a distribution of the places and of the capacities or the incapacities attached to these places. Put in other terms they are allegories of inequality, that is why you can change the value given to each term without changing the meaning of the oppositions themselves.
>
> (Rancière 204b: 7)

For Rancière, 'emancipation of the spectator' (and of theatre) 'starts from the principle of equality. It begins when we *dismiss* the opposition between looking and acting' on the understanding that 'the distribution of the visible itself is part of the configuration of domination and subjection' (2004b: 7, my emphasis). Equally, it begins with the dismissal of oppositions and hierarchies between mediums and forms such as theatre, television or literature and of equivalences such as that between theatre and community. Above all, it is based on a notion of a common power:

> [which is] the power of equality of intelligences. This power binds individuals together to the very extent that it keeps them apart from each other, able to weave with the same power their own way.... It is the capacity of the anonyms, the capacity which makes everybody equal to everybody. This capacity works through an unpredictable and irreducible play of associations and dissociations.
>
> (Rancière 2004b: 9-10)

Rancière asserts that, whether in a theatre, a museum, school, street or watching TV, 'there are only individuals weaving their own way in the forest of words, acts and things that stand in front of them or around them' (2004b: 9).

There is something emancipating about the idea of simply dismissing hierarchical oppositions. However, I'm not sure that at this point 'The Emancipated Spectator' doesn't fall back into the sort of much-criticized reversal of the historical production/reception model, such as that developed by Fiske in relation to TV under the influence of Barthes, Foucault and Althussar. Rancière also starts to suggest what 'theatre should do', something that is inconsistent with his argument in the same paper and elsewhere and that always sits badly coming from someone who (to my knowledge) has never 'done it'. Nevertheless, much of what he says chimes with something of my experience of *Susan and Darren* and perhaps of Doughty's and Mangan's of *EatEat*, although the terms of their argument differs. Within the space of their performances,

Quarantine's work seems to presuppose an 'equality of intelligences' of all concerned.

In respect of the performers, this presupposition is staged not just on the basis that 'ordinary people' like Susan participate in these shows speaking for and as themselves but that they do so in a fashion that 'holds open' the opposition between 'artist' and 'non-artist'. The claiming of the position of artist by those traditionally excluded from it by virtue of gender, class or ethnicity has been a major theme in performance concerned with identity politics since the 1960s and has often been discussed as part of a discourse of a troubling of the distinctions between art and everyday life. Nevertheless, while this gesture may have been accompanied by a rejection of traditional theatre spaces, subject matter, aesthetics and techniques, in taking up the position of 'artist' there is a *reification* of the gap between 'artist' and 'non-artist' and art and everyday life and a re-affirmation of a particular 'distribution of the places and of the capacities or the incapacities attached to these places' (Rancière 2004b: 7). In the context of experimental, performance venues and of *Susan and Darren*, Susan cannot be unambivalently identified either an 'artist' or a 'non-artist'. That is, there is no evidence that she identifies herself in these terms,[13] yet equally there is no practical distinction between her position in *Susan and Darren* and that of any other devisor/performer paid to participate in a show led by a director's artistic vision and concerned with high production values. In devised work, such performers also sometimes contribute deeply personal, even painful autobiographical material. I may have already noted a 'difference' from other such performers in the particular *quality* of her performance, but actually within the show itself there is no 'ground' to judge this quality as art (ifice) or not art (ifice). The identification of such ground would require the signalling of *either* a 'gap' or 'distance', or a lack of 'gap' or 'distance', between Susan and her appearance in *Susan and Darren*. There clearly must be a distance, since she is performing in an, at times, highly formalized spectacle but her representation in the piece offers no pointers for reading 'behind' her appearance.

This is partly because there is also no evident hierarchy of genres and mediums in its bricolage of the 'social', the 'popular' and the 'experimental', whereby one might be said to 'comment on' another. In short, while there may be intertextuality (something on which *all* meaning-making depends), there is no evidence of irony, parody, pastiche or other 'postmodern' modes of self-reflexivity and distanciation, designed to 'activate' the spectator and create a knowing distance from the spectacle and its subjects - Susan and Darren.[14] It is in the assumption and/or signifying of such a privileged 'knowing distance' and thereby potentially of an 'unknowing', even ignorant, subject/performer that issues of inequality, manipulation and voyeurism arise, whether in theatre, TV or elsewhere. On this point there are certain aesthetic similarities between *Susan and Darren* and some reality TV programmes hailed as being among the most ethical and affective of this 'genre', such as the documentaries of Molly Dineen or drama-documentaries of Paul Greenglass. These similarities include a tendency found in nineteenth-century naturalism, as defined by Emile Zola, to focus on the environment, or as TV critic John Caughie put it in the 1980s, they concentrate on the 'social space' in which the action takes place as opposed to the individual or 'character' (Caughie 1981: 243-6). As part of this, these programme-makers tend to dwell on detail, surface or appearances to an extent that, as with Zola's novels, their work becomes recognizable as high-*stylized*, creative exploitations of the medium of television. Which does not mean they necessarily draw attention to their own constructedness, any more than does any other work of art.

While not immediately apparent, within *Susan and Darren* the same stylization, and emphasis on theatrical appearances as evidenced in the pole dance and the washing scene, is given to all

[13] We asked Susan what had changed for her over the course of repeatedly performing the show. She told us that while she remained nervous before performances, as she became more accustomed to it she didn't see herself as performing at all but rather as having a conversation with people.

[14] Susan and Darren (and all of our other work) does use irony, but hopefully it is irony without cynicism.

Photo Simon Banham

other aspects of the piece. The show opened with a verbal sketch of the living room in Susan's house and with Darren moving around the space, fastidiously marking the size and placement of objects and furnishings and arguing with Susan about colour and texture. Certain objects became the springboard for anecdotes, yet the same meticulous, concrete description was given to everything in the room. Equally, in their monologues in 'the serious bit', Susan and Darren paid as much, indeed *more*, attention to the environment and material detail as to the individuals involved and to what might be thought of as the 'action' of these narratives. For example, before touching on his father's murder and Susan's rape, Darren spent some time delineating the exact geographical location of the piece of wasteland where these events took place (on the corner between Princess Road and Moss Lane East and bordered by Barnhill Street) and its position in relation to various clubs and drinking places (the Reno, The Nile, Shine's Shabeen and Barnhill Street Blues). Similarly, in recounting the rape he concentrated on details, such as how, as the rapist dragged Susan by her long red hair, she left a trail of coins falling out of the oversize pockets of her electric, blue dress.

What connects these monologues, the washing scene and the pole dance, therefore, is that we can empathize with the situations described and 'interpret' their emotions from what we see and hear. However, there is simultaneously too much and too little information to identify with Susan and Darren as the ground for social or political collectivity, or with them emotionally as the ground for community. Clearly we can classify Susan and Darren through discourses of gender, class, sexuality and ethnicity, but not only does this ignore the fact that they are *performing* but that on the basis of their appearance in this spectacle their identities are constructed as much, if not more, through regionality and locality, streets, clubs, specific rooms and objects, clothing, hair, profession, familial relationships and friendships, small habits, large random events, appalling co-incidences etc. In short, we encounter them as 'nothing more the infinite and self-evident multiplicity of human kind' (Badiou 2001: 25-6). This is that which *potentially* maybe allows for disruption of the police order but is in not essentially in and of itself either the basis of politics or of community.

Throughout the piece, then, we are simply presented with appearances in the presupposition that we will make sense of them, but it is *our* sense not Quarantine's or Susan's and Darren's. There is therefore a 'distance' from *Susan and Darren* and its 'subjects', but rather than a 'remarked' and 'knowing one', or one of 'mastery', or as something to be overcome or transcended, it is that which is 'the normal condition of any communication', which is *always* 'mediated' (Rancière 2004b: 6). It was this that gave the impression of being taken to the 'centre' of Susan's and Darren's lives but without privileged knowledge or access beyond that of anyone who might encounter them at their home or at a party. As such, paradoxically, it is the focus on surface, 'show' or appearances rather than what is 'behind' them and indeed 'behind' the show as a whole, socially, politically, personally or emotionally, that gives a sense of an 'authentic' encounter with Susan and Darren 'as one speaking subject with another'. What makes this encounter with Susan and Darren possible, but both separates and joins us in, as Doughty and Mangan put it, 'a metaphysic of unity based on separation', is quite simply *Susan and Darren*, the spectacle itself (Doughty and Mangan 2004: 38).

I have to admit that it is possible that the audience-participation and various breakings of the barriers between stage and auditorium in the show were in fact attempts to 'activate' the spectator and create a sense of community.[15]

[15] We feel that something of the distance of spectacle is removed by the audience's active relationship with the performance (conversation, making sandwiches, dancing etc.) and also somehow a shift for those audience members sat in rows away from the 'action', witnessing others more actively involved. This change in relationship, with Susan in particular, perhaps alters the sense of 'privilege'.

However, in the context of the lack of self-reflexivity, foregrounding or 'comment' in the piece, this felt to me like being offered a choice of perspectives in which no particular distinction was made between 'looking' and 'joining in' or 'acting'. In sum, it touched on the possibility of changing locations and occupations between performers and non-performers, artist and non-artists.

As some of Richard's and Renny's footnotes indicate, I am, of course, very much 'weaving my own way' through *Susan and Darren*, as I am through Rancière (Rancière 2004b: 9). Moreover, I am not claiming that this show produces a major reconfiguration of the sensible in theatre (let alone in 'politics'). Indeed, in direct contradiction to my comments on the staging above, within the performance I saw there was evidence of the way a particular distribution of the sensible still 'policed' what was doable for the spectator. On entering the space I was disappointed not to grab one of the 'comfy' armchairs in front of the video monitors but as the show progressed became relieved that I had not done so. Most of the action took place behind them, so that those sitting in them 'had to' twist round in their seats uncomfortably, or rather some unwritten 'rule' prevented them from simply picking up the chairs and turning them round.[16] Does this contradict most of what I have said so far? Only if we assume a 'regime' whereby a performance can or should function to exemplify particular theories without any sense of the excess that is always part of any genre of spectacle.

Further, as well as questions of 'community' in their article on *EatEat*, Doughty and Mangan are similarly concerned with politeness and the 'rules' governing audience behaviour (2004: 37–8). For me, the way issues of 'classification', order, place, occupation and genre gently accrue around Quarantine's practice, do signal at least some sort of 'disruption', if only of some now all-too-familiar presumptions about the 'ethical' and the 'political' effects of aesthetics in contemporary performance.[17]

REFERENCES

Aristotle (1974) 'The Art of Poetry' in *Aristotle, Horace, Longinus*, trans. H. Dorsch, Middlesex: Penguin.

Badiou, Alain (2001) *Ethics: an Essay on the Understanding of Evil*, trans. P. Hallward, London and New York: Verso.

Bondebjerg, Ib (1996) 'Public Discourse/Private Fascination: Hybridisation in True-Life-Story Genres', *Media Culture Society* 18: 27–45.

Caughie, John (1981) 'Progressive Television and Documentary Drama', in T. Bennett (ed) *Popular Television and Film*, London: British Film Institute.

Doughty, Sally and Mangan, Mick (2004) 'A Theatre of Civility', *Performance Research (*On Civility) 9.4: 30–40.

Fiske, John (1999) *Television Culture*, London and New York: Routledge.

Hallward, Peter (2006) 'Staging Equality: On Rancière's Theatrocracy', *New Left Review* 37: online <http://www.newleftreview.org/2view=2601>, accessed January 2008.

Kershaw, Baz (1992) *The Politics of Performance: Radical Theatre as Cultural Intervention*, London and New York: Routledge.

Laclau, Ernesto (1996) *Emancipation(s)*, London and New York: Verso.

Quarantine (2008) online <http://www.qtine.com/>, accessed between January 2007 and January 2008.

Rancière, Jacques (1999) *Disagreement: Politics and Philosophy*, trans. J. Rose, Minneapolis: University of Minnesota Press.

Rancière, Jacques (2004a) *The Politics of Aesthetics: The Distribution of the Sensible*, trans. G. Rockhill, London: Continuum.

Rancière, Jacques (2004b) 'The Emancipated Spectator', online <http://Rancière.blogspot.com/2007/09/Rancière-emancipated -spectator.html>, accessed September 2007.

Sedgwick, Eve Kosofsky (2003) *Touching, Feeling: Affect, Pedagogy, Performativity*, Durham and London: Duke University Press.

Skeggs, Beverly (1997) *Formations of Class and Gender: Becoming Respectable*, London: Sage.

Skeggs, Beverly (2004) *Class, Self, Culture*, London and New York: Routledge.

16 Those pesky armchairs. Always a problem. We proposed losing them entirely for the first tour. We felt that the audience occupying them were neither here nor there (if you know what we mean) and there was something folksy and fake about them as objects (old granny's chairs), that belonged neither in Susan's and Darren's home nor in our theatre space. But when we took them out we had this horrible, clichéd, 'multi-media performance' bank of monitors staring back that was way too present. So the armchairs stayed, with some regret.

17 Last words. In the final performance at Contact on the second tour, Darren had begun the 'Living Room' section when the back doors of the studio were flung open (not the doors that any of the rest of the audience had entered through). In strode two large middle-aged women. Darren turned and without missing a beat, said 'Oh come in Dotty, Marie... Grab yourselves a chair, you haven't missed much, I've just been describing our living room'. So Dotty and Marie (old friends of Susan and Darren) grabbed two of the armchairs and stuck them on the dance floor in prime position at the edge of (but clearly 'on') the stage. Susan and Darren (show and performers) continued with Dotty and Marie solving our armchair problem.

Intensities of Appearance

ADRIAN KEAR

> - Appearance is not an illusion that is opposed to the real; it is the introduction of a visible into the field of experience, which then modifies the regime of the visible (Rancière 1999: 99).

In *Disagreement* (1999: 74), Jacques Rancière identifies two distinct modalities through which appearance emerges within, and operates upon, what he terms 'the distribution of the sensible': the essentially political organisation of sense-making activities and apparatuses within the established framework of the intelligible and the visible (*essentially political* because any 'distribution' involves the allocation of parts and division of participatory opportunities; *essentially political* because it then differentiates between agency, activity and perspective on the basis of position and role rather than equality and capacity; *essentially political* because in the end 'politics is a question of aesthetics, a matter of appearances'. The first of these, he claims, operates through the deployment of 'a singular mimesis' (1999: 74) facilitating the appearance of a homology between 'the people' as political constituency and people as material reality; between 'the people' as represented as either multitude or coherent community and people's presence as destabilising multiplicity - 'always too numerous or too few compared to the form of their manifestation' (1995: 96). This mechanism is not a matter of simple resemblance (*mimesis* in the restricted sense) but of a certain 'regime of resemblance' (*mimesis* as a political framework) ensuring 'resemblances function within a set of relations between ways of making, modes of speech, forms of visibility and protocols of intelligibility' (2007a: 73). This, he explains, is what enables the appearance of the political within a given historical situation or 'distribution': the mobilisation of 'the *appearance* deployed by the name of the people' in a particular political direction or configuration. (Rancière 1995: 96; emphasis in original). For Rancière,

> Politics exists ... because there are names which deploy the sphere of appearance of the people, even if in the process such names are apt to become separated from things; ... and because the name of the people is at one and the same time the name of the community and name of a part of - or rather a split in - the community. The gap between the people as a community and the people as division is the site of a fundamental grievance (1995: 96-97).

It follows that the second modality of appearance emerges from, and operates within, this 'site', providing 'a singular disruption of this order of distribution of bodies as a community' through the appearance of 'what comes and *interrupts* the smooth working of this order through a singular mechanism of subjectification' (1999: 99; emphasis added). In other words, appearance appears at a moment of re-animation, and re-imagination, of genuinely political action within the current political 'distribution'.

Appearance is thereby situated as both a mechanism for producing and maintaining political 'consensus' and a vehicle for 'staging specific scenes of dissensus' (2004: 304) otherwise foreclosed by 'consensual' practices

Performance Research 13(4), pp.16-24 © Taylor & Francis Ltd 2008
DOI: 10.1080/13528160902875598

aimed at suturing and stabilising the gaps, fissures and fundamental incommensurability between the appearance of the political and the appearance of politics. The 'singular mechanism of subjectification' referred to above is thereby conceived as political subjectification; or, rather, subjectification in and through the re-appearance of the possibility of politics. Dissensual practices, according to Rancière, create 'a modification of the co-ordinates of the sensible' (2007b: 259) and an opening for political subjectification precisely by re-opening the division, the foundational separation, between presence and representation - between the claim to inclusive political community and the concomitant exclusion that inaugurates and grounds its intractable division. For Rancière,

> Politics exists when the natural order of domination is interrupted by the institution of a part of those that have no part. This institution is the whole of politics as a specific form of connection. It defines the common of the community as a political community, in other words, as divided, as based on a wrong that escapes the arithmetic of exchange and reparation. Beyond this set-up there is no politics. There is only the order of domination and the disorder of revolt (1999: 11-12).

This article aims to approach the above as a methodological imperative as much as a philosophical position: to foreground particular loci of interruption and moments of modification in order to identify the specifically aesthetic-political dynamics of practices of 'dissensus', and to investigate the challenges made to 'the distribution of the sensible' and the 'regime of the visible' through the negotiation of particular modes of democratic participation and image production. It takes as its primary focus a number of works by the photographer and video-artist, Phil Collins, which provide an exemplary exposition of the relations between presentation and representation operative in the dynamics of appearance and its alteration, and offer something of a case-study in their aesthetic-political intensification.

WOUNDS TO THE FACE

> The democratic experience is thus one of a particular aesthetic of politics. The democratic man is a being who speaks, which is also to say a poetic being, a being capable of embracing a distance between words and things which is not deception, not trickery, but humanity; a being capable of embracing the unreality of representation. A poetic virtue, then, and a virtue grounded in trust
> (Rancière 1995: 52).

In May 2003, the Iranian political dissident, poet and human rights activist, Abbas Amini, initiated a remarkably simple, silent intervention into the institutional and cultural politics of asylum in contemporary Europe. Taking a household needle and coarse green thread, he sewed together his lips, ear lobes and eyelids in direct, visceral response to the UK Home Office's 'inhumane' decision to appeal his successful application for indefinite leave to remain. At first this action appeared intended as a private gesture of defiance, hidden from public view with the complicity of friends; but the economy and vitality of its exposure of the 'public secret' of the British state's derogation of responsibility for the rights of the refugee meant that, once made visible, Abbas Amini's action quickly became 'an international cause celebre' and his situation instantly newsworthy (*The Guardian*, 31 May 2003). Although by no means unique - a group of Iraqi detainees at the Woomera Camp in South Australia had similarly sutured their lips in protest at their treatment as 'illegal immigrants' the year previously, and the practice certainly extends beyond these examples - the appearance of his name, and more tellingly, his face, brought into focus the violent effects of the discourse of asylum and provided the locus for a necessary *interruption* of their deployment in the public policy debates and governmental strategies of Western nations. Amini swiftly recognised the international political significance of his act of localised resistance and re-inflected his one-man hunger strike into a global demand for equality and justice, claiming: 'I have sewed my eyes so that others could see, I have sewed my ears so that others could hear, I have sewed my lips to give others a voice' (BBC News, 30 May 2003). Despite the fact that fairly early on in the case the Secretary of State was refused the right to appeal the tribunal's decision, thereby removing the impending threat of deportation back to Iran and the resumption of his life in prison, Amini continued his self-violating exercise in sense-deprivation for eleven days and nights with the expressly *political* intention of drawing attention to the interests 'of refugees everywhere' (*The Guardian*, 31 May 2003). His commitment to continuing his 'tortured gesture of agency', to use Jospeh Pugliese's fine phrase (2002; cited in Gilbert and Lo 2007: 189), with not inconsiderable personal pain and suffering, bespeaks both fidelity to the task of representing the incessant impact of torture and dehumanisation and, at the same time, acute awareness of his status as a political subject with the capacity for staging what Rancière calls 'scenes of dissensus' within the playing space of his own embodied self-presentation (2004: 304).

The critical moment in the above action appears to have come with its decisive entry into the field of vision, its condensation into the appearance of an *image*, captured unerringly by Phil Collins in his remarkable photographic portrait, *abbas amini* (2003).

The work reveals not only the clear discomfort caused by the heavy-handed needlepoint and the barely concealed anguish that led to the cotton thread's insertion; it evidences as well the intellectual clarity, composure, fortitude and

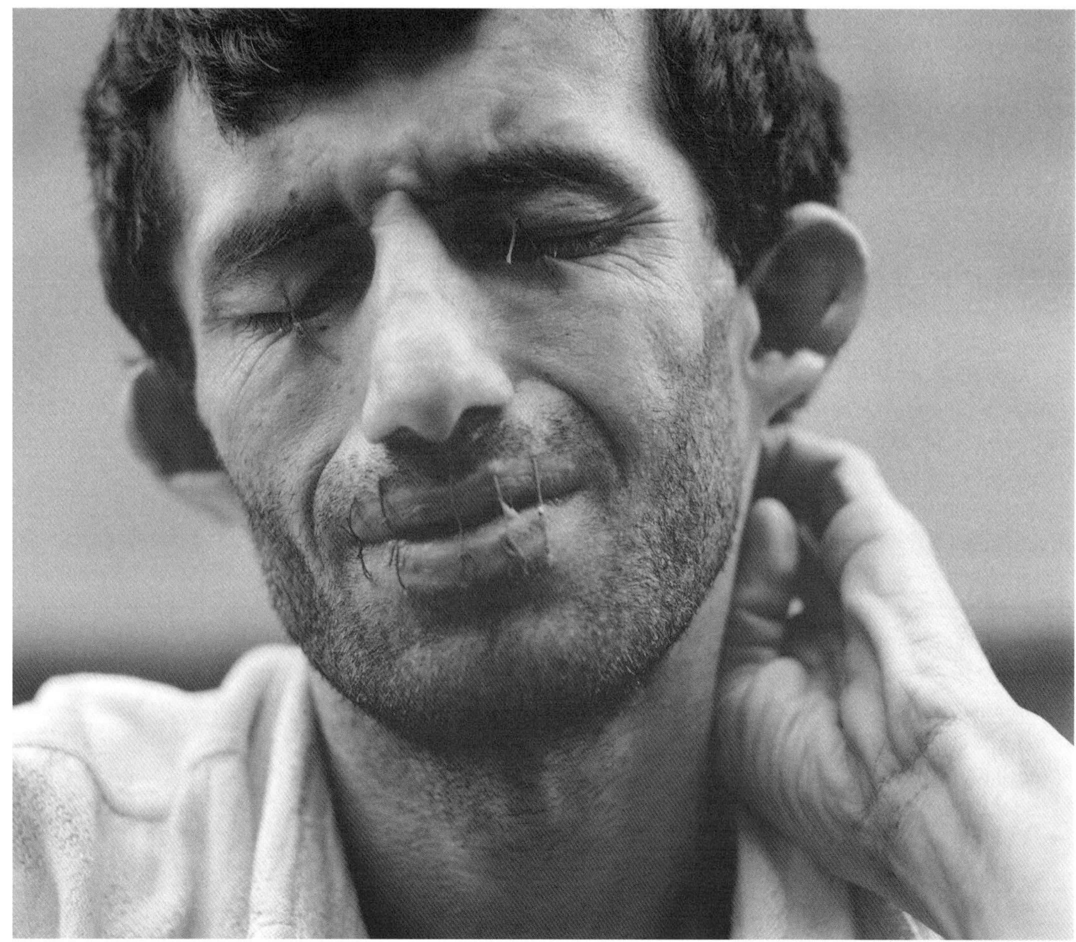

• **Phil Collins, *abbas amini*, 2003.** *Lightjet print on Fuji Crystal Archive paper reverse mounted under Diasec, 60 x 70 cm. Courtesy the artist, Kerlin Gallery & Tanya Bonakdar Gallery © Phil Collins*

rigour underlying its subject's radical gesture of resistance - a defiance of de-subjectification which might be expressly regarded a politically subjectivating act in its own right. Jenny Edkins, following Agamben, has called this moment of re-subjectification through de-subjectification, '*the assumption of bare life*, that is, taking on the very form sovereign power seeks to impose' (2005: 3; emphasis in original) and thereby instantiating a relational logic of resistance. This relationality is inscribed in the structure of the image itself, implicating its subject, maker and spectators as co-responsible mediators of the political situation and context which it frames and represents. In contrast to Bourriaud's valorisation of the instrumentalised relationality the aesthetic event (1998), Collins draws attention to the intrinsically political relations at stake within the aesthetic formation of the dissensual, interruptive event. Amini's eyes appear to stare inwards from beneath their lids, drawing the viewer outwards from the image itself to the occasion, and significance, of its making; at the same time they seem cast downwards as the spectator is pulled into an interior space of subjective destitution and its defiant assumption. In this respect, the photograph seems to be a re-presentation, or 'reading', of an event which already consciously stages the material presence of the body as a site of visible resistance; it's affective address functioning through the correspondence between Amini's facial expressiveness and Collins' artistry and technical prowess. The designation of the title of the work in lower case letters - a feature of all Collins' projects - operates in recognition of the fact that what it appears to make visible is the transformation of an individual subject into the embodiment of politically constituted processes of subjectification and de-subjectification, with the framed and focused *image* bearing the apparent imprint of their visceral effect. Moreover, its explicitly *theatrical* quality suggests that the work is invested in laying bare the mimetic dynamics of the discourses and relations of power and violence which inscribe it, exposing not only their manifest materiality but also their 'hidden dependence on illusion' (Taussig 1999: 54). Collins' portrait reinforces the recognition that Amini's self-presentational action is precisely an intervention into the field of *representation* - a performance of strategic 'defacement', to use Taussig's particularly apposite term, which usefully draws attention to the corporeal dynamics of the ideological suturing of reality to appearance (1993: 233). By creating a negative image of theatre as 'the space of visibility of speech, the space of problematic translations of what is said into what is seen' (Rancière 2007a: 88), Collins' portrait captures the core theatricality of Amini's political gesture with remarkable precision and clarity.

Importantly, then, the photograph cited here must be seen as a collaboration between two agents actively participating in bringing it into existence: Amini sews up his eyes in order to produce the affective demand to be recognised, and seals his lips in order to claim the voice that was properly his to begin with; Collins demonstrates the interpenetration of the sayable and the visible in an image whose explicit 'content' concerns the inaccessibility of both under certain conditions of appearance. In the image, Amini's ostensibly self-inflicted wounds to the face seem to reference not only the veiled scars of suffering that might reside elsewhere on the body - in jail in Iran he had his feet and extremities beaten repetitively, his arms shackled and body hung above the ground for days on end with regularity (BBC News, 27 May 2003) - but to directly correlate them with the current situation in Britain. As an intervention, the image appears designed less to substantiate Amini's case for asylum by graphically illustrating his familiarity with pain and violence than to provide an indication of their interrelation in subjectification, an interpretation of their political linkage and interpenetration. Whilst Amini's action seems to link two distinct events otherwise separated by the distance of time as well as geography, and in the process to mark a point of continuity between them, its condensation in the photographic image also appears to suggest a rupture, a break, in the semblance of continuity and apparent coherence offered by the discontinuously experienced, and practiced, hegemonic framework of 'universal human rights and justice'. Even as Amini's stitches hold in tension - and temporary suspension - the desire for personal testimony alongside political commitment to the demand of equality, and articulate a coextensive call for both recognition and a say in the 'democratic' processes that generated their necessity, the photograph which shows this most clearly also invites the spectator to encounter Amini's presence within it as the presentation of presence *as such* - as the appearance in the form of the image of something otherwise occluded from appearing within the existing regime of the sayable, the visible, and the sensible. Through its framing of Amini's face it radiates an intensified apprehension of the interconnection between representation and presentation, between the elements of a situation and the logic governing the conditions of their appearing, between who and what 'counts' - and why - and who and what does not.

Yet at the same time as offering us this 'staging', at the same time as demonstrating the visceral dynamics of subjectification and de-subjectification, the assumption of bare life and its political intensification, the image precisely does not give us Abbas Amini (upper case) himself: he remains secreted, veiled, closed to view in the work that bares both his name and face: remaindered from representation even as his appearance marks its interruption. What subsists in the image is, rather, the artistic presentation of his presence as such: Collins' theatrical re-presentation of the face-to-face relation as precisely a relation facing us; a representation of the political relation par excellence (Rancière 2007a: 23, 15).

• **Phil Collins, *they shoot horses*, 2004.** *Installation view, Phil Collins, Neue Kunsthalle St. Gallen, St. Gallen, 2006. Courtesy the artist & Neue Kunsthalle St. Gallen © Phil Collins*

• **Phil Collins, *they shoot horses*, 2004.** *Synchronised two–channel colour video projection with audio, 2 x 420 min. Courtesy the artist, Kerlin Gallery & Tanya Bonakdar Gallery. © Phil Collins*

VISIBLE RETURNS

> The task of politics is to return appearance itself to appearance, to cause appearance itself to appear (Agamben 2000: 94).

In *they shoot horses* (2004), a split screen video-art installation filmed in Ramallah and shortlisted for the 2006 Turner Prize, Phil Collins continues to place the appearance of ordinary people at the heart of his aesthetic project. The work consists of two single-camera videos, shot over a seven-hour duration, projected onto two differently sized and perpendicularly placed screens.

The videos capture the spontaneous performances generated by two sets of young Palestinians in response to Collins' 'set-up': a day-long non-stop disco dance marathon underscored by instantly recognisable Western pop. The young performers simply move, without direction or any other form of authorial intervention, within the frame of the image established by the fixed camera lens and the strong horizontal plane created by the parallel lines of the pair of orange stripes on the wall and the dance-hall's wooden floor. The shallow depth of focus and narrow stage-space contributes dramatically to the distinctive frontality of the performance, with the performers positioned in a fixed relationship to the static camera even as their movement appears to exceed what can be contained within its frame. It is the framework of representation, in other words, that governs and conditions their presence - or, rather, their *appearance* - within the presentational dynamics of the work. The effect of this is redoubled in the environment of the installation by the impact of the sheer size of the two sets of synchronised projection, with the floor to ceiling aspect ratio of the right-hand one in particular creating the impression of a larger-than-life face-to-face relation. And yet, despite the proximity generated by the performative frontality, the distancing effect generated by the materiality of the image itself - ontological status as image, rather than presence as such - serves as a reminder that there is no face-to-face relation in the context of mediation, just the bare flicker of its remainder supplementing representation's perpetual presentation of the relation of non-relation.

In other words, the face-to-face situation established by Phil Collins, as artist-maker, in Ramallah, is rendered, through the dynamics of mediation and spectatorship in the remove of the gallery context, as potentially exploitative as the participative dance marathons referenced intertextually through the work's title (see *They Shoot Horses, Don't They?* [dir. Sydney Pollack, 1969] which trades on the expropriation of popular participation into a commercial entertainment function). This, after all, is how the apparatus of representation - and media representation in particular - tends to work: through the distanciation of abstraction, objectification, and de-contextualisation. But at the same time, and with equal clarity and force, the work constantly reminds us there is never simply representation without presentation, never pure performance without the destabilising presence of the performer themselves.

Over the course of the piece, by means of its seven-hour duration, what we witness is something like the appearance of this logic - and thereby something of the logic of appearance itself. In their enthusiasm, elation, extroversion and determination, we glimpse a little of the young performers' individuality, their creativity, their agency; and through their exposure, their enjoyment, their desire to dance and the inevitable exhaustion we see an index of their fecundity, fragility, and fundamental humanity. We see them in a light unfamiliar from mainstream media representations, a light whose intensity enables their strange familiarity to shine through the rigid containment and pictorial distance of the screen/scene. Yet at the same time we also see the constant presence, and routine intrusion, of the social situation in the material arrangement of the performance itself. The continuity established by the spatial frame of the image and the constant variation of movement within it is interrupted at several points by power cuts, technical cock-ups and the calls to prayer of an adjacent mosque. These moments of disturbance act as a pertinent reminder that the conditions governing the appearance of the young people within this context are at least as determined by the social situation as by the aesthetic regime of representation. In this respect it's perhaps worth noting that, almost as though to make the point, the Israeli border police chose to confiscate the tapes containing the second hour of Collins' recordings as he exited Ramallah. Yet Collins' creative intelligence lies in being able to anticipate and intentionalise such recuperative dynamics within his artistic technique. His dogged adherence to duration, and to the discipline of the framing, serves to ensure that the apparently naïve actions of the performers evade censorial attention, at least in part because there appears too little of note in their performances to mention. In its simplicity, its radical economy, the work presents an *excess* of presence over the regime of representation even though the form such presentation takes is limited to the appearance of 'mere appearance' within it.

Collins seems well aware of this when, right at the end of *they shoot horses*, he moves the camera for the first - and only - time in the piece, zooming-in to create a close-up of one of the young men. Helen Moseworth draws attention to this dramatic alteration in a recent article for Artforum, describing it as 'an astonishing moment' of rupture in which the camera 'practically caresses the intensely beautiful, flushed and exhausted face' (2008: 236). However, rather than seeing it as a failure on Collins' part - a an eruption of his desire into the space of otherness; a breach of the discipline the work had established - we might instead see it as a calculated comment designed to draw attention to his continuous presence, controlling the camera; a moment of exception revealing both the working of the mechanism and the labour of making. Collins is clearly aware of the significance of the deliberate disruption as a coda for artistic disclosure. He notes in interview that:

> I've always seen the person behind the camera as a central element of the equation. ... The witness to these very fragile and beautiful moments, the person who sits and takes the picture, is very much implicated - the one who directly influences the activity (2007: 86).

The movement of the lens that focuses on the young man's appearance is then, in this context, a movement that marks the return to visibility of the 'triangular structure made up of artist, performer and audience' (Kotz 2007: 64). The effect is to underpin the differential relations at play - and at stake - in the creation of the work and the recognition of its situatedness.

THINKING THE SITUATION

> Appearance is nothing but the logic of a situation, which is always, in its being, *this* situation. Logic as a science restores the logic of appearance as the theory of situational cohesion in general
> (Badiou 2004: 172; emphasis in original).

In the recent work of Alain Badiou, the concept of *appearance* performs a similar function to that of 'the distribution of the sensible' in Rancière: it codifies both the organising logic of the situation and the specific forms of its manifestation. For Badiou, appearance is the mechanism which connects, disconnects and re-connects the elements of a situation in a particular configuration, accounting for both the specificity of their conjunction - their 'distribution', *pace* Rancière - and their intrinsic volatility and variability. Appearance works, it would seem, through the generation and realisation of different degrees of intensity; where intensities of appearance provide a means of codifying a complex network of relationality. The logic of appearance is therefore primarily concerned with the relation between the elements of a situation and the conditions governing their mode of appearing, between their ontological presentation and its orchestration in a structured and cohesive regime of representation. Importantly, this is never fixed and final, for at the same time as granting the relationship between the elements of a situation apparent consistency and temporary stability, the logic of appearance leaves open the possibility of an alteration in the intensity of their appearing, an 'unbinding' of their relationship, even the 'local collapse' and 'temporary cancellation' of its logic in the form of the transformative singularity of the *event* (Badiou 2004: 175).

It might be suggested that such abstract philosophical formulations have little direct bearing on the aesthetic choices and tactical interventions made by practising artists thinking through the dynamics of concrete situations and material relations. But perhaps the essentially political thinking of the situation apparent in both forms of intellectual practice provides sufficient justification for their juxtaposition. For Badiou, 'the essence of appearing is the relation. ... Appearing ... is what superimposes the world of the relation onto ontological disconnection' (2006: 162). In other words, appearance grounds the relationship between the elements of a situation - ontological presence - and their organisation and inter-relation in the structure of representation through creating localised 'consistency' in the relational intensities through which they are presented. Appearance is therefore the representational form of the logic of the situation thought in terms of its apparent cohesion (2006: 164). Politics exists, for Badiou, in the appearance of modes of practice, 'diagonal to representations', which proceed through 'the material critique of all forms of presentative correlation' - the very logic of appearance which aims 'to prevent their appearance' as such (2005: 77). These practices seek to effect the 'unbinding' of the relations between the elements of a situation; an alteration of the relative intensities with which they appear; a 'redistribution' of parts, positions and visible participants. Perhaps Collins' art practice can be seen, and read, in this context. The photographic work, *abbas amini* (2003), appears to capture a moment of 'refusal of relation' (Bayly 2007: unpaginated correspondence) undertaken by an agent whose chosen form of self-presentation - sensory denial and self-mutilation - makes manifest the relations of power and (in)visibility operative within the situation and reduplicates their effects through their performative 'assumption'. The artist's work here lies in exposing dynamics otherwise displaced from visibility, facilitating the intensified appearance of a face (and voice) otherwise rendered inapparent in the normalised relations of representation. In the video-installation project, *they shoot horses* (2006), Collins' concerns are likewise directed towards making a work in which the normally unrepresented elements of a political situation - the energy, creativity and vitality of ordinary young Palestinians - are presented with an aesthetic intensity which draws attention to their apparent invisibility. At the same time, the framing of the piece, and the context of its installation in the gallery, works to 'unbind' the spectatorial relation anticipated and engaged by its mode of performative self-presentation in order to create a disjunctive *theatrical* experience of the relation of non-relation which, for Badiou - and, it would seem, Collins - is the essence of genuine politics.

ACKNOWLEDGEMENTS

The author gratefully acknowledges the generosity of Phil Collins in granting permission to reproduce the images included in this article. Thanks are also due to colleagues on the 'Intensities of Appearance' panel at PSi 13, *Happening, Performance, Event*, 8-11 November 2007, New York University (Simon Bayly, Gianna Bouchard, Joe Kelleher, Alan Read, and Janelle Reinelt); and to Helen Gilbert and Jenny Edkins for sharing their thoughts about lip-sewing as a form of protest. Carl Lavery kindly commented on the article in draft form and contributed several helpful suggestions. All errors and oversights are, of course, the sole responsibility of the author.

REFERENCES

Agamben, Giorgio (2000) *Means Without End: Notes on Politics*, trans. Vicenzo Binetti and Cesare Casarino, Minnesota: University of Minnesota Press.

Badiou, Alain (2004) *Theoretical Writings*, ed. and trans. Ray Brassier and Alberto Toscano, London and New York: Continuum.

Badiou, Alain (2005) *Metapolitics*, trans. Jason Barker, London: Verso.

Badiou, Alain (2006) *Briefings on Existence: A Short Treatise on Transitory Ontology*, trans. Norman Madarasz, New York: SUNY Press.

Bayly, Simon (2007) 'On Appearance', unpublished notes and correspondence with the author, 25 October 2007.

Bourriaud, Nicolas (1998) *Relational Aesthetics*, Paris: Les Presse du Reel.

Collins, Phil (2007) 'Don't Blow Your Own Horn' (interview), in *Phil Collins: the world won't listen*, ed. Suzanne Weaver and Sinisa Mitrovic, Dallas Museum of Art; New Haven and London: Yale University Press, 2007, pp. 85-98.

Edkins, Jenny and Véronique Pin-Fat (2005) 'Through the wire: Relations of power and relations of violence', *Millennium Journal of International Studies*, vol. 34, no. 1: 1-24.

Gilbert, Helen, and Lo, Jacqueline (2007) *Performance and Cosmopolitics*, London: Palgrave.

Kotz, Liz (2007) 'Live Through This', in *Phil Collins: the world won't listen*, ed. Suzanne Weaver and Sinisa Mitrovic, Dallas Museum of Art; New Haven and London: Yale University Press, 2007, pp. 57-65.

Moseworth, Helen (2008) 'Man with a movie camera', *Artforum*, January 2008, pp. 232-239.

Rancière, Jacques (1995) *On the Shores of Politics*, trans. Liz Heron, London: Verso.

Rancière, Jacques (1999) *Disagreement*, trans. Julie Rose, Minneapolis: University of Minnesota Press.

Rancière, Jacques (2004) 'Who is the Subject of the Rights of Man?', *South Atlantic Quarterly*, 103: 297-310.

Rancière, Jacques (2007a) *The Future of the Image*, trans. Gregory Elliott, London: Verso.

Rancière, Jacques (2007b) 'Art of the Possible: Fluvia Carnevale and John Kelsey in Conversation with Jacques Rancière' in 'Regime Change: Jacques Rancière and Contemporary Art', *Artforum International*, March 2007, pp. 252-285.

Taussig, Michael (1993) *Mimesis and Alterity: A Particular History of the Senses*, London and New York: Routledge.

Taussig, Michael (1999) *Defacement: Public Secrecy and the Labor of the Negative*, Stanford, CA: Stanford University Press.

Figuring the Face

SIMON BAYLY

• Figure 1. The masks of comedy and tragedy.

Let's face it: this is a bad start, this sign is not a good sign. As the all too familiar icon for anything and everything theatrical, the masks of comedy and tragedy, or the *personae*, appear to have degraded into crudely line-drawn caricatures, the grotesque grimaces of an inane grin and an utterly artificial sadness. In fact, this is worse than bad: this is embarrassing.

Martin Heidegger recalls the etymology of persona as the actor's mask in one of his lectures collected in the series published as *What is Called Thinking?* After declaring that man is 'the animal that confronts face-to-face', he adds: 'since man is the percipient who perceives what is, we can think of him as the *persona*, the mask, of being' (Levin 1999: 282; emphasis in original). But not even the philosophical gravitas of Heideggerian thinking can prevent an instinctive revulsion towards this symbol of the *personae*. One finds it on supermarket noticeboard adverts for amateur dramatic events, decorating the garish facades of some commercial theatres, lurking as historical relics in passageways of more modern venues, even in corporate graphics for projects in affective computing, but not usually, say, in the promotional literature for university courses in the study of performance. There is more to this revulsion than distaste for artless graphic design. In the forms of representation these masks have taken, is there not simply ugliness, but also falsity, pretence, dissimulation, even something diabolical?

If only it were so interesting. In fact, does this symbol not announce, as those who have throughout history either condemned or ignored it are fond of asserting, theatre's essential moral sickness as a form of playacting that is merely a debased, rather than a demonic, infantility? Should not any form of intellectual enquiry be utterly ashamed of this idiotic symbol, shamed *by* it and also wishing shame *upon* it so that it would hide its ridiculous, gurning faces forever? Again, let's face it: who goes to the theatre anyway? Aren't many of us - even those who profess theatre itself - in an unspeakable but tacit agreement with the good company of Roland Barthes when he remarks: 'I've always liked the theatre, and yet I hardly go there anymore' (Barthes in Scheie 2000: 162).[1] Probably with good justification, the philosophically inclined do not generally regard the theatre as a good night out - the statistical chances of the stirring of desire or curiosity are low. Not even the attempted intellectual rehabilitation of theatre by the exhortations of Alain Badiou or the more restrained considerations of mimesis explored by Phillipe Lacoue-Labarthe would appear to have

[1] Barthes' development of a corporeal poetics in relation to the repudiation of his former passion for theatre is excellently elaborated in Scheie (2000).

Performance Research 13(4), pp.25-37 © Taylor & Francis Ltd 2008
DOI: 10.1080/13528160902875606

persuaded contemporary philosophy otherwise (Badiou 1990 and Lacoue-Labarthe 1998). While the scholarship of theatre and performance has been profoundly affected by 'theory' (and the post-theory 'turn'), the theatre is fondly, politely but quickly forgotten, even as philosophy decks itself out in the language of performance in a proliferation of theoretical scenes, stagings and acting out. No doubt, this is a characteristic exaggeration, itself symptomatic of a secret passion for histrionics, which, when its passions are ignited (all too easy) cannot stop hoping to attract attention by some self-aggrandisingly rhetorical self-depreciation. To be properly creditable, pretence should come pseudo-po-faced, bearing an expression of insufficiently disguised distaste that is itself insufficiently disguised. This is pretence to the nth power, where n is as high an integer as you can fake it and still look scrutably inscrutable, a facial manoeuvre that has left the building before the body even knows it.

Within a Judeo-Christian heritage, the shameful symbol of the personae commits the ultimate travesty of erasing the difference between masks and faces. Christianity has conferred an entirely negative value on the mask. As Michael Indergand suggests in his inspirational survey of the grimace, if humankind is indeed made in God's image, then it must follow that everything that serves to dissimulate or disfigure it must be the work of the devil, a manifestation of the folly of humanity in the grip of spiritual malaise (Indergand 1982). In other religions too, a properly spiritual countenance is a picture of serenity: is the limpid face of the Buddha not in some way the same as that imprinted on, for example, the Turin shroud? But it suffices to be reminded that the latter is a face of a dead man, or at least a great, fake impression of one. The life behind it, or immanent in it, has vanished. And the Buddha seems barely alive: he is without desire and his symmetrical face remains empty, devoid of expression, the eyes merely depressions in the stone. Only his hands, often bent at the wrist in anatomically impossible angles, hint at the operation of an autonomous intelligence, but one that is situated in an ineffable 'elsewhere' (Siegel 1999: 102). But of course, the Buddha is not so easily corralled within such a deadpan physiognomy - hence his common figuration as the 'laughing Buddha', the one who mocks his own serious pretensions.

On the other hand, depictions of devils across the gamut of cultures show faces deformed and distorted, shadowy, craggy, hairy, wrinkled, bedecked with horns and misshapen ears, gigantic noses, wearing leering grins or snarling with teeth exposed. Their faces are violently animated as well as animalistic, almost unable to contain the expressive forces of malevolence that play in and out of their convoluted surfaces. The malificence of masking, its roots in shamanic ritual and its connection to the supernatural is well established in many cultures and has a long and rich history in ethology and anthropology. But if God is dead, then so is the devil and all things devilish become merely silly theatrical contrivance: voodoo kitsch all dressed up with nowhere to go.

At a secular level, the distinction between mask and face has been reconstituted as the paradox of acting. That paradox (or, rather, that pretence at paradox which assumes a degree of complication that may not exist) suggests that the actor, rather than personally or internally undergoing the emotions that they represent, requires only the skilled and conscious deployment of the behaviour or signs that correlate to the relevant emotions. In other words, when it comes to a truly convincing performance, going through the motions will be more successful than undergoing the emotions. Denis Diderot's well-known essay on this subject, written between 1773 and 1778, obviously stands as a pivotal work in the history of this paradox, but it exercised the minds of many of those who thought about theatre before and after its dissemination, and still holds sway over, for example, competing conceptions of acting technique. In 1888, William Archer, unhappy with Diderot's cold-hearted dictums, published his own investigations into the paradox. In a work distilled from biographical and journalistic research into actors' own sense of their craft and entitled *Masks or Faces?* (Diderot and Archer, 1957), Archer's title literally focuses the paradox upon the human face, where it has remained in the subsequent elucidation of human emotion in philosophy, psychology, and in the rapidly developing science of the interface, or human-computer relations.

However, if the theories of subjectivity that have emerged over the last 30 to 40 years have been seeking a refuge from the logic of representation or the ideology of the gaze, then it is *the body* that has provided the most capacious hideout in this respect. The recalcitrant, resistant and abject body, subjected to all manner of penetrations, dismemberments and interpretations, has become the resting place for many approaches seeking to establish the importance of qualities of experience and the transmission of knowledges that are not regulated by the aggressive exercise of cognitive rationality. Contemporary theory has insisted upon a reconsideration of the body as theme in great depth, following in the wake of a philosophical strain of thought best evoked by Nietzsche in his typically trenchant attack 'On the Despisers of the Body':

> 'Body am I, and soul' - thus speaks the child. And why should one not speak like children?
>
> But the awakened and knowing say: body am I entirely, and nothing else; and soul is only a word for something about the body.
>
> The body is a great reason, a plurality with one sense, a war and a peace, a herd and a shepherd. An instrument of your body is also your little reason, my brother, which you call 'spirit' - a little instrument and toy of your great reason [...].
>
> Behind your thoughts and feelings, my brother, there stands a mighty ruler, an unknown sage - whose name is self. In your body he dwells; he is your body.
>
> There is more reason in your body than in your best wisdom (Nietzsche 1954: 146).

Yet, despite the ubiquity of *the body* within contemporary academic discourses, the philosophies of embodiment often seem curiously bereft of any differentiation of what is often vaguely described as 'felt, bodily experience' to the extent that it seems that we genuinely lack an appropriate language with which to explore the intricacies of the relation between *psyche* and *soma*.

Perhaps the relative neglect of the face as a necessary part of a phenomenological analysis of embodiment is partly to do with its attributed complicity with a 'metaphysics of presence'. In this view, the face as a figure of presence or 'the clearest expression of the soul' would stand as visual counterpart to the primacy given to the speaking voice as bearer of expressive authenticity. Just as this primacy has assiduously been unravelled by Derridean deconstruction as an essentialising phoncentrism, definitive of an autonomous, self-hearing speaking subject posited by the philosophies of modernity, so the thinking of looking and being looked at has, since Sartre's lengthy chapter on 'the look' in *Being and Nothingness* (Sartre 1969: 252-301), been preoccupied with the scopic desire of the gaze and the optics of surveillance as elaborated in the discourses of feminism and film theory.

Yet despite the critical successes of deconstruction, or perhaps because of them, it increasingly seems that there might be other dimensions to be recovered from the face as a phenomenon that does not operate purely to shore up a dessicated analysis of the metaphysics of presence, the self-possession of the transcendental ego or a patriarchal ideology of objectification. Such a project is central to the work of perhaps the most influential philosopher of the face, Emmanuel Lévinas. To the extent that theatre itself is predicated on a face-to-face encounter, on a mutual presencing (however much haunted by spectral absences), it would seem particularly suited to Lévinas' construal of 'ethics as first philosophy' directly out of a prolonged consideration of the 'face-to-face'. In relation to a theatrical understanding, Lévinas' phenomenology of the face is well worth a detailed and nuanced examination. But what is relevant here is that Lévinas' thinking of the face resolutely objects to any direct phenomenological consideration of the face in its appearing. For, to put it simply, in Lévinas' metaphysical phenomenology, the face is not a phenomenon; it cannot appear. The face just faces. It does not represent and cannot be represented without a fundamental betrayal or violence. The face is everywhere as a force of radical passivity in Lévinas' texts, yet nowhere described, differentiated, articulated, examined or scrutinised either as an active power or as a coded surface:

> The face is meaning all by itself. You are you. In this sense, one can say that the face is not 'seen'. It is what cannot become a content which your thought would embrace; it is uncontainable, it leads you beyond. It is in this that the signification of the face makes its escape from being, as a correlate of knowing. Vision, to the contrary, is a search for adequation; it is what par excellence absorbs being. But the relation to the face is straightway ethical (Lévinas 1985: 86–87).

Elsewhere he maintains:

> The face is present in its refusal to be contained. In this sense it cannot be comprehended, that is encompassed. It is neither seen nor touched - for in visual or tactile sensation the identity of the I envelops the alterity of the object, which becomes precisely a content (87).

Thus, for Lévinas, the face is not given as set of data to perception: the sum of the eyes slightly narrowed, the brow wrinkled, the lip curled, the corners of the mouth raised or depressed. Neither is it a 'gestalt', a composite figure against a ground, nor a succession of expressions geared around the production of an intentionality. The face simply faces, addresses, commands. To the extent that the face surrenders to its phenomenality, it surrenders its ethical significance; it becomes just 'a face', an image or representation, a set of features for conveying information.

But Lévinas is equally clear that this ethical dimension 'opens in the sensible appearance of the face'. How are we then to understand the way in which the face and facing *in experience* have the primordial significance that Lévinas attributes to them? David Levin has made an astute analysis of the problematic of this invisible face of humanity in Lévinas' work and of the ambiguities and paradoxes of a phenomenology of the face which is not a phenomenon. In seeking a phenomenological narrative for the face, reduced in the later work to no more than 'the trace of a trace of an abandon' (Levin 1999: 93) that Lévinas does not supply, Levin asks us to think that the obligation in the address of the face:

> first takes hold of us bodily - in the flesh - in a time that is, at each and every moment, i.e., both synchronically and diachronically, prior to thematising consciousness, prior to reflective cognition, and therefore prior to the ego's construction of a worldly temporal order ... morality is first of all a bodily carried sense of obligation, an imperative sense of responsibility immediately, but not consciously felt in the flesh: a bodily responsiveness that, unless severely damaged by the brutality of early life experiences, the 'I' cannot avoid undergoing - at least to some extent - when face-to-face with the other (279).

Here, the undifferentiated body-as-flesh is once more invoked as the inchoate site of an ethical signification beyond what is available to knowledge. Is this body any more 'phenomenal' than the vision of a face that is no more vision, but listening and word? In attempting to articulate this sensibility, Levin invokes a peculiar reading of Lévinas' project:

> For him, the formation of the moral self involves tearing off the masks, returning to one's exposedness, making felt contact with that existential condition and living from out of that, without the mediation of the masks. From an ego-logical point of view, this exposure of the face to the face of the other would be the unspeakable terror of self-effacement, the most extreme deconstruction of the identity of the 'self' as the culture of modernity has conceived it. There is no

> telling what identity-shattering effect this exposure of the face behind the mask could have on the eyewitness, the one who sees it. (281-282).

For Levin, the mask hides and - even as it indicates what it hides - betrays the face, which ought to remain as the name for what is hidden, made neither into image nor symbol.

In this understanding, the mask (and our newly recuperated symbol of the personae) is simply a disgrace, a grotesque rendering of what ought to remain unrendered but nevertheless felt or otherwise experienced. Put bluntly, I think this is a profound but useful misreading. What a theatrical thinking offers in response is that the function of the mask is not to hide the face as site of pure revelation of the individual soul but rather to reveal it as a complicated and complicating figure of appearance, or even:

> the appearance of appearance, the figure of figuration, the ur-appearance, if you will, of secrecy itself as the primordial act of presencing ... a contingency, at the magical crossroads of mask and window to the soul, one of the better-kept public secrets essential to everyday life (Taussig 1999: 30).

In erasing the distinction between mask and face, the graphic symbol of the personae announces the impossibility of a pure appearing or a simple revelation of selfhood. In other words, facing is unthinkable without masking, voicing without ventriloquism. But while such ideas have become familiar critical currency, I am also grasping for another understanding, a thinking that takes up this symbol of what might be called a philosophical physiology of the composure and de-composure of being: the all-too-human microprocesses of becoming unaccommodated, undone, overwhelmed, confounded or otherwise losing control. This would mean bringing a grown-up and self-confident rationality into immediate proximity with the derangements of the infant, the animal and the senile. This is what the theatre can show us in a singular way - though it rarely does or even desires so - and what a theatre-philosophy can think.

THE FACE: WHAT A HORROR! [2]

If I had to conjure a replacement icon for the *personae* to represent such a theatre-philosophy invoked earlier, then I would be hard pressed to find another as richly complex and unhappily beautiful as this:

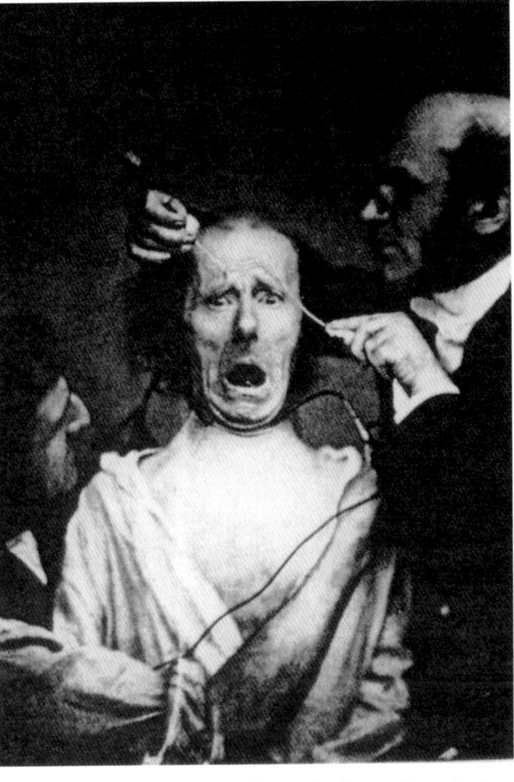

• Figure 2. Plate illustrating the emotion of 'terror' in Duchenne's *Mechanism of Human Facial Expression*.

The scene depicts, without doubt, a disquieting event: a dramatic instant of expression both violent and ambiguous. What is the proper emotional response to such an image, evidently staged for our benefit? Perhaps an instinctive mirroring of the fear (or is it terror?) expressly inscribed on the chalky white face of the Christ-like innocent at the centre of the photograph, surround by the darker forces of God knows what. But, then again, is this event a tragedy or a comedy? Should one laugh or cry? Or permit one's interest to be led by the gaze, mobilising the critical faculty of thought, or rather, out of respect, feign disinterest? Or look away in disgust or avert one's eyes out of shame? And is not the cause of our confusion precisely the issue

[2] I take my cue here from the exhortation, 'A horror story, the face is a horror story' in Gilles Deleuze and Félix Guattari, *A Thousand Plateaus: Capitalism and Schizophrenia* (Deleuze and Guattari 1988: 168).

of *resemblance*, or lack of it, at the heart of expression? For, in the shadow of the preceding discussion, what this image appears to express is the fear (which is also a form of excitation) that not only does 'what is expressed' bear no resemblance to the expression (fear of what? - what is there to be afraid of in this picture, or just beyond the margins of its frame?) but also that it no longer even relates essentially to what expresses itself, to the figure whose face bears an expression that does not seem to belong to him. If so, we might conclude that what appears in the instant fixed by this image is the very horror of the 'abstract machine of faciality' (Deleuze and Guattari 1988: 168) itself which has demonically installed itself in the body of a human being and is making its face express - what?

Is not this image a sublime allegory of expression, an iconic repository of its immanent force, which expresses more than we might ever say or write about expression; a picture that, for once, might indeed be worth many thousands of words? In his meditation on the uncanny effects of masking and unmasking, Michael Taussig asks: 'could the face, as both window and mask, ur-border and mother of all borders, be allegory, thus defined?' In a reading of Walter Benjamin on allegory, Taussig goes on to suggest that the allegorical object is 'incapable of emanating any meaning or significance of its own; such significance as it has, it acquires from the allegorist. He is ... within it and stands behind it not in a psychological but in an ontological sense ... and it becomes for him a key to the realm of hidden knowledge' (Taussig 1999: 252-253).

As an experimental scene that is also, as Barthes says of the photographic image, a primitive theatre, this particular facial staging seems to engage Benjamin's notion of allegory to the letter (Barthes 1981: 31). For this photograph's *raison d'être* is not simply as the record of an experimental procedure, clearly showing an allegorist literally - as well as psychologically or ontologically - at work, but as the very evidence of a truth that is *the inscription of the theatrical*, of the representation of that which exceeds representation. But perhaps this is already to read far too much into what is after all only one image among an infinity and into what merely appears to be written on the face of its central protagonist. In making a crisis out of the drama of this particular image, do we give too much credit to the power of the abstract machine of faciality in completing its imaging of the world?

EXPRESSIVE MECHANISMS AND MECHANICAL EXPRESSIONS

In the 1840s at the massive Salpêtrière hospital-cum-human dumping ground on the outskirts of Paris, the French physician Guillaume Duchenne de Boulogne began to use a large but portable battery to apply electric current to the bodies of patients suffering a variety of physiological and psychological conditions, with apparently surprising and reportedly beneficial effects. During the course of his treatment, Duchenne observed that electrical stimulation of particular muscles always provoked specific and regular physical behaviour - the jerking of a limb in the same way, or the twitching of an eyebrow. It was, above all, the animation of the facial muscles that caught Duchenne's scientific interest and also his aesthetic ambition. After making little progress in mapping the then poorly understood nature of the facial musculature and nervous system using the heads of the recently deceased or guillotined, Duchenne came across an elderly man who suffered from almost complete facial anaesthesia - making him an ideal subject for the experimental project. It appears that Duchenne was able to modulate his electrical apparatus so that he could later use the faces of fully sensitive subjects, without inflicting excessively unpleasant sensations upon them.

Duchenne presented the results of his experiments in 1862 with the publication of the *Mechanism of Human Facial Expression*, a work that has had a profound effect on the study of emotion and facial expression from then on (Duchenne, 1990). The book was a landmark publication in another way, in that it required the

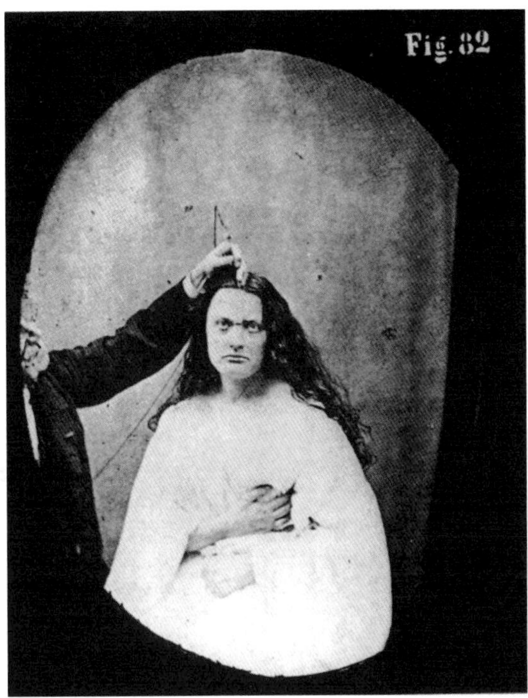

• **Figure 3.** Plates 82 and 83 from ***Mechanism of Human Facial Expression.***

innovations not just of electricity to produce its findings, but of another then emerging technology to present them: photography. Unimpressed by the initial efforts of professionals and their failure to understand the physiological facts, Duchenne taught himself the requisite photographic skills. By use of electrical stimulation, Duchenne was able to 'freeze' expression on the faces of his subjects for the several seconds required by contemporary photographic technology, thus fixing what had been noted for millenia as all too swift and transient. The result was a series of extraordinary images reproduced in one of the first publications containing photographs, rather than drawings or engravings - marking the beginnings of a process whereby the image no longer functions merely as illustration but as *evidence* - a form of objective, scientifically valid proof.

Duchenne envisaged his project as a ground-breaking contribution not only to anatomy and physiology, but also to psychology, philosophy and the fine arts. The album is divided into two separate sections, one 'scientific' and the other 'aesthetic'. In the former, the face is divided according to a system of one-to-one correlations between particular facial muscles and specific emotions. For each muscle/emotion pair, Duchenne presents an image of his experimental subjects undergoing localised electrical stimulation of that muscle. Thus we have images for the muscles of attention, reflection, aggression, pain, joy and benevolence, lasciviousness, sadness, weeping and whimpering, fright and terror, and so forth.

In the 'aesthetic' section, Duchenne 'corrects' the facial expression of various Greek and Roman sculptures, deemed deficient in their truthfulness due to the artists' poor understanding of the physiology of facial expression. More adventurously, he also stages some electrophysiological scenes of emotion, many of which are produced literally as dramatic excerpts, adding costume and gesture to the stimulation of facial expression.

Plates 82 and 83 (figure 3) depict one of Duchenne's blind female patients receiving electrical stimulus to the 'muscle of aggression'. The images are described in the text as explicit stagings of moments from Shakespeare's *Macbeth*, for example:

> Plate 82: Lady Macbeth: Had he not resembled
> My father as he slept, I had done't.
> Moderate expression of cruelty. Feeble electrical contraction of m.procereus [...]
> Plate 83: Lady Macbeth - about to assassinate King Duncan. Expression of ferocious cruelty. Maximal electrical contraction of the m.procerus (122).

Duchenne reveals the extent of his dramatic imagination by adding:

> We see that these darkened features are singularly ugly. I have imagined that Lady Macbeth, in recognizing a resemblance between King Duncan and her sleeping father, lost her courage to strike and collapsed onto a seat. (This scene is not in Shakespeare.) (122).

Theatre, and photography understood as a form of theatre, is thus at the heart of Duchenne's experimental project. The most extreme 'staging' of the face in these images occurs when Duchenne divides the face of his performers in two in order to demonstrate the general modifying effect of one expressive muscle on all the other facial features. The reader is then invited to experience the drama of the scene by covering first one half of the face, and then the other, so as to perceive two differing emotional states.

For example, the legend for Plate 79 (figure 4) reads:

> *Maternal happiness mixed with pain*, from a psychological and aesthetic study of the expression resulting from the *conflict* of *joy* and of *crying*.
>
> By covering the left eye, joy of a mother who sees her infant recovering from a serious illness; covering the right eye, the same maternal joy, united with pain produced by the death of another child. Electrical stimulation of *m. corrugator supercilii*, associated with the natural expression of joy (204; emphasis in original).

Here, rather than the features of the face as the clearest expression of the soul, the face is the site of the struggle for expression between opposing forces of the psyche:

> We have seen, in Plate 82, the expression of cruelty that I have given her, under the impression of her

• **Figure 4. Plates 78 and 79 from *Mechanism of Human Facial Expression.***

> horrible invocation. But quite at home in the art of dissimulation, Lady Macbeth, in going to receive her king, has already been able to dissipate the last traces of her furious transport of anger, and her mouth composes itself into a false smile, as she knows how to reinforce the affectionate and lying remarks.
>
> Here, happily, the control of her face ceases. She cannot impose on her gaze a sympathetic expression in harmony with her smile [...]
>
> [...] Her mouth is smiling in saying 'Your servants ever.' But what a smile (cover the right side of the mouth and of the cheek)! See how the eye is cold and freezes the smile. It is a smile that kills (127).

Duchenne's claims for the 'idealised realism' of his aesthetic studies would seem to be merely curious from a contemporary perspective, if it not were for the overtones of torture, technophilia and the mad, bad scientist about his work - something that seems to resurface whenever performance and electricity come into direct association. However, the theatricality for which Duchenne appears to have had an obvious flair is as integral to his 'scientific' images as it is to their 'aesthetic' counterparts. The uneasy combination of artist, scientist and photographer played by Duchenne is perhaps as compelling to the contemporary eye as it was abhorrent to many critics of the time. In their explicit obsession with artifice, intervention and invention, Duchenne's activities appear as anathema to conventional scientific method to the extent that they stand as a properly theatricalised experiment, in which evidence is all too clearly and literally solicited, manipulated and distorted.

In the typical nineteenth-century photographic portrait, the space of the photograph is often literally converted into theatrical space, complete with artificial scenery, drapes, props and costume. Or, in the more classically inspired work of portraitists such as Julia Margaret Cameron, the context in which the sitter is placed is carefully rendered as provisional, abstract or elusive, offering up a 'there' which refers to nowhere. All signs are removed that the scene is also a site of photographic work that is instead is rendered as a fictive place which it is tempting to interpret as an analogue to the bourgeois idea of the discrete, autonomous, self-governing individual. But the thinking of theatre I am attempting here can only consider such an idea - an abstract bourgeois 'then' when the people that mattered supposedly believed in and, more importantly, *lived* as discrete, autonomous, self-governing individuals - as itself a necessary fiction. This fiction is what stares out at us from the faces of those Duchenne animates in his experiments in images where the wires *are* literally showing, where the scene as a site of work is palpably explicit.

Whilst Duchenne was unhappy about the presence of himself, his hands and electrodes in the images, it is just this presence that renders them devoid of the kind of theatricality we can retrospectively see at work in early photographic portraiture. His technique requires a tactile contact between experimenter and subject, a literal completion of the circuit that will make expression take place and take shape. His hands and electrodes are as much part of the elicited expression as the skin and muscles of the subject's face: both are part of the apparatus, which is not simply reducible to its technological elements. Most of the images in the 100 or so handmade copies of the album are severely cropped versions of the originals - whose reproductions are apparently still used in undergraduate anatomy classes at the Ecole des Beaux Arts in Paris. In the uncropped images, the entire experimental scene is depicted, including the electrical apparatus and the dark-suited Duchenne and his assistant working with apparently intense concentration - the lighting and composition given as much attention as the placing of the electrodes. The cropping of the images for publication was presumably to focus the reader's attention on the subject to hand, the painting of 'the expressive lines of the emotions of the soul on the face of man' (Duchenne, G-B., et al. 1990: 9). Yet

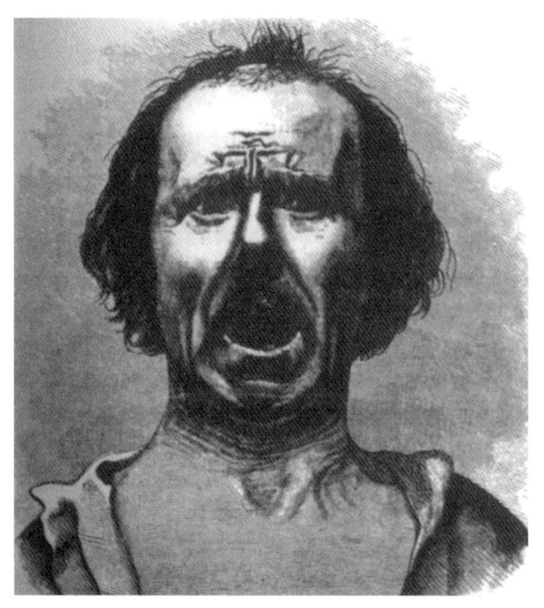

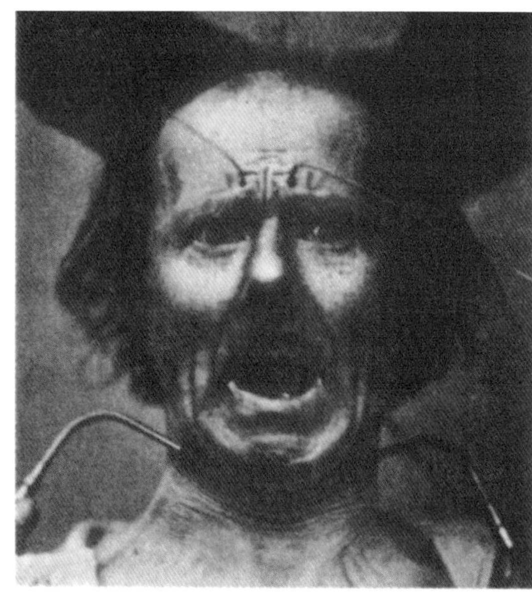

• **Figure 5. Darwin's engraved reproduction from *Expression of Emotions in Man and Animals* alongside Duchenne's original photograph illustrating the expression of terror.**

Duchenne and the experimental set-up remain almost laughably visible - there is just no way that the experimenter can erase his performance. To do so would mean that the entire scene and event of expression would have to be redrawn and thus de-authenticated as evidence - in much the same way as anatomical dissection destroyed what it attempted to reveal of the facial mask of musculature. This is precisely what Charles Darwin did in his borrowing of Duchenne's images for his *Expression of Emotions in Man and Animals*, published some 10 years after Duchenne's Album (figure 5).

Where Darwin's strategy is to classify and illustrate emotional expression so as to demonstrate what is properly and universally human, Duchenne's album displays a curious human puppetry in which an interior Cartesian theatre of the mind, in which I might be said to 'own' my own expressions at a conscious or unconscious level, is turned inside out and made into a theatrical system of connected individuals and technologies, channelling and capturing the forces of electricity and expression.[3]

Duchenne's images stand as both an unintentional critique and a literal *mockery* of the conventional theatricalised portrait. There is no illusion of inwardness created on behalf of the subject, certainly no self-expression; neither is there post-modern masquerade, a play with identity or the manipulation of a role. What is given to be seen here is a different kind of event, closer to what Hegel describes as 'the sense of a reality separated and cut off from the individual outer expression, in which the individual no longer retains possession of himself per se, but lets the inner get right outside him, and surrenders it to something else' (MacIntyre 1972: 88). Indeed, the acts that Duchenne depicts are separated and cut off from the individuals undergoing them, since they are inscribed upon the face by externally stimulated muscle contraction. But something beyond Hegel's phenomenology is also expressed: the sense that 'I' am not the one letting the inner get right outside, nor surrendering it to something else, but rather that *it is happening in me*, or, better, that 'I am' such a happening, and one that determines the fluctuating limit of the self-knowledge of a self-conscious rational agent.

Looking at Duchenne's images today, they appear not as expressions of emotion at all, but rather their opposite: like the *personae* with which we started, these are grimaces, grim refusals of expression, fixed at the moment of their artificial excitation, the album as a whole, a

[3] I owe this notion, as well as much insight into the representational strategies of medical illustration, to Philip Prodger's remarks on Darwin's use of illustration in 'Photography and The Expression of the Emotion', in Darwin et al., 1998.

graphology of defacement. Experienced this way, the images present neither information nor evidence, but questions addressed directly to the viewer as interlocutor, questions that appear to emanate from the voiceless face of the one who is being defaced at the centre of each image: *Who is doing this to me? In whose name is this being done to me? What is happening? What are you looking at?* These effects are compounded by Duchenne's use of the series in producing the differing emotions on the same subject in a similar experimental and photographic set-up; the old man is run through the gamut of facial expressions, his face subjected to a lifetime of emotion, the extremes of which perhaps he, like the rest of us, barely approached in life with any degree of awareness.

In an opposed fashion, the contemporary art of the photographic portrait generally prefers to present faces as solitary enigmas; perhaps most clearly exemplified by the early work of Thomas Ruff or Wolfgang Tilmans, its ethical responsibility is to respect the integrity of its subject - the individual - in the act of exposure, however extreme or gratuitous. While various inflections of theatricality have become a well-established feature of contemporary art photography, from the Philip-Lorca diCorcia's 1990 *Streetwork* series to the work of a younger generation of female photographers such as Anna Gaskell and Justine Kurland, as Alan Trachtenberg remarks, one has to look to the domain of the amateur, to the pornographic or domestic snapshot (usually part of a series) or home video to find a different kind of 'realist' presentation of self and others, event and context, organised around a real or hastily imagined narrative of some kind: a party, a beach holiday, a birth (Trachteneerg, 2000: 4). Typically, such images often show their 'subjects' half-heartedly gesturing towards a theatrical or cinematic frame of reference, demonstrating a sense that some kind of performance 'for camera' is indeed required but for which it is hard to summon the requisite enthusiasm or energy.

But in the series of images that depict him, Duchenne's old man has no context, narrative or role of his own: he simply appears, bare of pretext. Yet we see his face apparently in the throes of terror and transformed by joy, in mild dejection and moderate good humour, in expressions of feeling that he neither initiates nor imitates. Through Duchenne's unconnected stagings, the young blind woman appears utterly removed from her various roles as Lady Macbeth, saint in ecstasy, mother in turmoil or coquette. Despite our knowledge of the fact that the expressions they wear are not their own, as individuals they become familiar: our familiars or representatives. This is reinforced by the appearance of other faces - a young girl, an opium addict, an older woman - in the series; after their anonymous, anomalous interventions, we return to the faces we have apparently come to know. And we ask another question: who are they?

Yet, despite superficial appearances, Duchenne is not Dr. Frankenstein: he is not making life where there is none or none meant to be. His electrodes extend the path of the facial nerves outwards into the environment by direct contact, exteriorising the normal flow of the impulse between the internal nervous and muscular systems, making explicit what is inherently implicit: he wants to show what a face can do. But the agency of its making and unmaking appears immanent only to the theatricalised world of the experimental system, rather than to any of its specific elements, human or otherwise.

The sense of theatre I am ex-pressing out of Duchenne's album here is all the more marked in contrast to the literal presence of theatre therein. In addition to the decidedly crude spectacle of the young blind woman put through various bit parts, a *bona fide* actor also appears in Duchenne's album in five plates, 'an artist of talent and at the same time an anatomist', chosen specifically because after much practice, he could 'by calling on his feelings ... produce perfectly most of the expressions portrayed by each of the muscles of the eyebrow' (figure 6). But even Duchenne's actor-anatomist could only manage to control one of his eyebrows, a deficiency he made up for by conforming to 'the requirements of physical beauty ... evident in his portrait' where, according to Duchenne, 'in repose, his features are handsome and regular' (Duchenne 1990: 43-44).

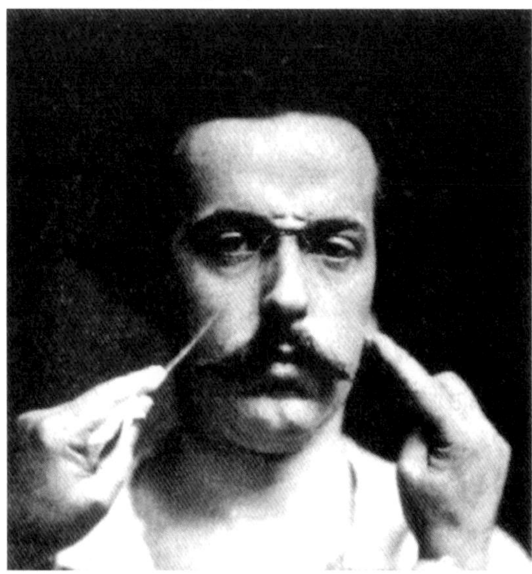

• Figure 6. Plate 24 from *The Mechanism of Human Facial Expression.*

Here, the actor, who to modern eyes is surely a picture of foppish affectation is presented as the gold standard of an authentic faciality *vis-à-vis* the degraded and passive faces of the neurologically impaired. Strange how, in the sense of theatre I am seeking here, he is the least theatrical figure in the album.

Lest we think we are here penetrating the mysteries of expression in overcoming the infantile tendencies of theatrical convention, Duchenne plays a trick on us.

The image pasted into the frontispiece of the Album (figure 7) depicts Duchenne side-by-side with the old man, the electrical apparatus and electrodes in full view applied to the latter's face, apparently inducing a disconcerting smile that resembles nothing more than the spasmic grin fixed on the mask of comedy. At first glance, Duchenne himself seems to take up the tragic mask of expression, or at least an appropriately dispassionate air. But looking again, the old man's smile vaguely resurfaces upon Duchenne's face, as if formed there by some unforeseen electrical feedback. There's something happening in Duchenne's self-presentation; a moment of transition. The unhappy ethics of experiment submerged in the drama of the Album 'proper' here appear apparently unadorned; the fake, artificial grin of the submissive old man - known in today's psychology as the Duchenne smile - and, the scientist's own modest expression, evidencing a natural and authentic good humour. Yet, as an aside, much later in the text, which like this one really plays a minor supporting role in relation to the pictures, Duchenne lets us in on a secret:

> Frontispiece A to this volume illustrates the method of electrization [sic] that I have used to obtain an isolated contraction of the facial muscles. The electrodes, held in my right hand, communicate with my induction apparatus via some conducting wires and are positioned to stimulate the muscles of joy. The expressive lines of joy would have appeared on the face of the subject if I had current through my apparatus. But I must say that in this case the laughter is natural! I merely wanted to show a

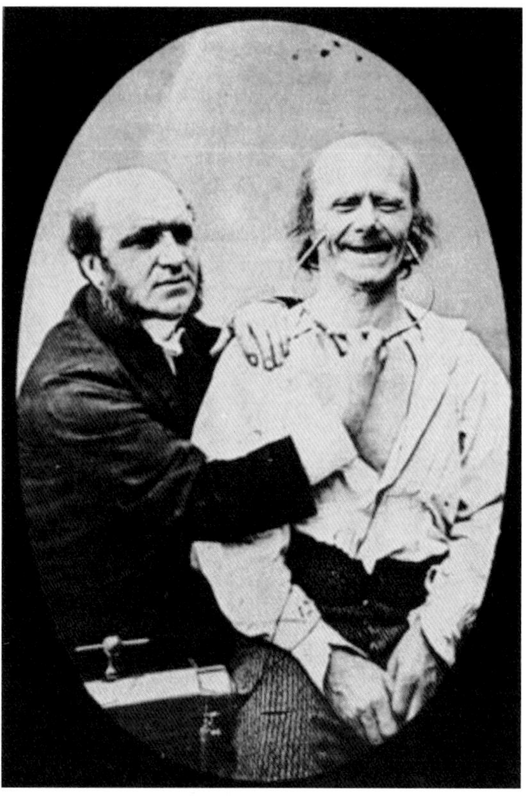

• Figure 7. The frontispiece to *The Mechanism of Human Facial Expression.*

simulation of one of my electrophysiological experiments in this figure (44, emphasis added).

It's a set-up: the image is a joke. Within its conceit, is Duchenne trying his best to keep a straight face? What 'fronts' the Album - its face, so to speak - is the representation of a simulation of an artificial stimulation of an artificial expression of joy representing natural joy. How does one begin to untangle the theatrical machinations going on here? The last laugh must be that immanent in Duchenne's own expression, masked (but not hidden) behind the front cover of his own picture book, where we are almost bound to overlook it in the desire to plunge directly into what we presume is the authentic heart of the matter.

REFERENCES

Badiou, Alain (1990) *Rhapsodie pour le Théâtre*, Paris: Imprimeries Nationales.

Barthes, Roland (1981) *Camera Lucida: Reflections on Photography*, trans. Richard Howard, New York: Hill and Wang.

Darwin, Charles, et al. (1872/1998) *The Expression of the Emotions in Man and Animals*, 3rd edn, with an introduction, afterword and commentaries by Paul Ekman. London: HarperCollins.

Deleuze, Gilles & Guattari, Felix (1988) A *Thousand Plateaus: Capitalism and Schizophrenia*, trans. Brian Massumi, London: Athlone Press.

Diderot, Denis & Archer, William (1957) *The Paradox of Acting* [with William Archer's Masks or Faces?], New York: Hill & Wang.

Duchenne, G.-B. (1990) *The Mechanism of Human Facial Expression*, ed. and trans. by R. Andrew Cuthbertson, Cambridge: Cambridge University Press.

Indergand, Michel (1982) 'Grimace et Grimaciers', in Baudinet M.-J. & Schlatter, C. (eds.) *Du Visage*, Lille: Press Universitaires de Lille.

Lacoue-Labarthe, Phillipe (1998) *Typography: Mimesis, Philosophy, Politics*, trans. Christopher Fynsk, Stanford: Stanford University Press.

Levin, David (1999) *The Philosopher's Gaze: Modernity in the Shadows of Enlightenment*, Berkeley, London: University of California Press.

Lévinas, Emmanuel (1969) *Totality and Infinity: An Essay on Exteriority*, trans. Alphonso Lingis, Pittsburgh: Dusquesne University Press.

Lévinas, Emmanuel (1985) *Ethics and Infinity*, trans. Richard A. Cohen, Pittsburgh: Duquesne University Press.

MacIntyre, Alistair (1972) 'Hegel on Faces and Skulls', in MacIntyre, A., (ed.) *Hegel: A Collection of Critical Essays*, New York: Doubleday & Company.

Nietzsche, Friedrich. (1883-5/1954) 'Thus Spake Zarathustra', in *The Portable Nietzsche*, trans. and ed. Walter Kaufmann, New York: Viking.

Sartre, Jean-Paul (1969) *Being and Nothingness: An Essay on Phenomenological Ontology*, trans. Hans E. Barnes, London: Routledge.

Scheie, Timothy (2000) 'Performing Degree Zero: Barthes, Body, Theatre,' *Theatre Journal* 52: 161-181.

Siegel, James (1999) 'Georg Simmel Reappears: "The Aesthetic Significance of the Face"', *Diacritics* 29, 100-113.

Taussig, Michael (1999) *Defacement: Public Secrecy and the Labor of the Negative*, Stanford, Calif: Stanford University Press.

Trachtenberg, Alan (2000) 'Lincoln's Smile: Ambiguities of the Face in Photography', *Social Research* 67(1), 1-23.

Returning Appearance to Itself
Trisha Brown, Koosil-ja and the materiality of appearance

MUSETTA DURKEE

A performer appears onstage, her back turned towards the audience, and begins to dance. She moves constantly, her body past her youth, her shoulder blades and spine highlighted by the overhead stage lights. The mechanics of her sweeping movements are articulated by her skeletal back. She never turns to face the audience for the duration of the solo - they know her, not through her face or expression, but through her appearance onstage. When she finally faces the audience to bow, her face seems foreign and unfamiliar, as if being greeted by a stranger. Trisha Brown, in her solo *If You Couldn't See Me* (1994), is nothing other than appearance. By never yielding to the audience's objectifying gaze, Brown resists the audience's attempts to seize hold of her appearance, to separate her real, physiological, performing self from her image, or to objectify her performing self into a significatory product for consumption. Instead, Brown's solo embodies and enacts what Giorgio Agamben's claim that 'all living beings are in the open: they manifest themselves and shine in their appearance' (2000: 91). Freed from the audience's potential seizure of her image by means of an authoritative gaze and mechanisms of signification, Brown's appearance onstage does not represent an underlying subject nor does it reveal any underlying truth. Her anonymous performance 'does not tell the truth about this or that state of being, about this or that aspect of beings and of the world' (2000: 92); it is, instead, appearance itself appearing.

Fast forward twelve years and enter into a stage-space split into two with bleacher-seating on each side, back-to-back and live-feed cameras portraying the activities occurring on the opposite side on a large screen. Momentarily confused as to whose mediated-depiction is being viewed on the large screen, the audience is faced with a reflection of another audience's appearance and another audience's gaze. In Koosil-ja's *Dance Without Bodies* (2006), this transformed stage-space frustrates the audience's role as unobserved, authoritative viewer and the images of choreographer Koosil-ja and fellow dancer Melissa Guerrero's moving bodies, mediated by the live-feed cameras, cannot be anchored or seized in the process of performance. Koosil-ja and Guerrero, instead of overtly denying the audience's gaze as Brown does, complicate the audience's gaze by constantly changing sides of the stage-space and by frantically shifting their own gaze between five clusters of television sets set up on each side of the stage-space, almost too busy to be bothered with the onlookers trying desperately to catch their eye. The audience observes their moving bodies but, like the audience of *If You Couldn't See Me*, the audience for *Dance Without Bodies* does not enjoy the position of objective, singular and authoritative voyeur. Instead, Koosil-ja's use of new media to create an extended sensory system via random sequencing of images on the television screens which motivate the dancers' movements forces the audience to engage with the affective and performative elements of their appearance instead of with their possible

Performance Research 13(4), pp.38-47 © Taylor & Francis Ltd 2008
DOI: 10.1080/13528160902875614

significatory aspects. It forces, in other words, the appearances of the dancers, both in the stage-spaces and on the screens, to be understood as appearance itself instead of the sign, marker or representation of an underlying, fixed, stable object of their viewing.

Both Brown and Koosil-ja therefore challenge, in different ways, the audience's objectifying gaze and free their appearances in the stage-space from merely representing some underlying, concealed reality. The stakes in freeing their appearance from such captivity are high, especially in live performance and dance: in holding an image on a screen captive or even the images of actors portraying a character on the stage, the audience is in control of the performing subject's projected appearance, a mediated - either through the camera or the character itself - image; however, in holding the appearance of a moving body of a dancer who is playing no role and is presenting movement itself captive through its objectifying gaze, the audience has control over the performing subject's physiological, material body - the flesh and blood entity which bears and suffers the consequences of racial, ethnic, gendered, ageist persecution and discrimination. The need to 'return appearance itself to appearing' (2000: 95) - that is, to undo the audience's seizure of the body's appearance and allow the image to exist as appearance itself - is therefore both political in that hierarchical dominance of material, flesh and blood bodies is a political dominance of oppression, control and even violence, as well as theoretical in that the material body challenges human spiritual ideals of physical transcendence and is thereby expunged into the realm of the nonhuman.

The task of causing appearance itself to appear is therefore a political imperative as well as an aesthetic one: taking hold of, possessing and arresting the representations of bodies, of persons, is an authoritarian act of domination and oppression. In performance, the moving body is often objectified, turned into an object for the consumption of the audience. Causing appearance itself to appear, as Agamben's project suggests, requires not only complicating the hierarchical distinction between the real and appearance and representation's role in mediating between appearance and reality, but also complicating the distinction between representing subject and represented object that such a hierarchy presupposes. Without challenging the ontological and political hierarchies of appearance versus materiality and without challenging the representing subject and its relation to the object of appearance via acts of representation and signification, appearance is doomed to remain 'a signifying proposition of sorts' (2000: 92).

My aim for this paper, then, is twofold: first, to free appearance from its relegation as 'mere' projection, signification, representation of an underlying 'real'. This task involves challenging the objectifying gaze of the audience which seizes hold of the dancers' bodies as images, denies their potential for pure communicability - that is, communicating the very act of communication as opposed to communicating, or conveying, the meaning of an underlying object - and thereby forces them to appear as evidence of some meaning or underlying object instead of as appearance itself. And secondly, my aim is to insist, as a footnote to Agamben's argument, that the material, moving, decaying flesh and blood body *must*, for both political and theoretical purposes, be incorporated into this act of pure communicability, of appearance itself appearing. This latter task involves understanding the relationship between the material body and its appearance without resorting to defining the material body as the underlying reality from which its appearance emanates and instead understanding appearance itself appearing in the reposed state of dialectical struggle - in 'the open' (2000: 91) - between appearance itself and the material, physiological and political reality of the body. To tackle this difficult task, I will turn to Agamben's discussion of Walter Benjamin's 'saved night', but I would like to first discuss how the dominance of the audience's gaze in live performance reifies and enacts the hierarchical binary between both authoritative subject and viewed object as well as between, what Bertrand Russell defines as, 'what is' and 'what seems to be'.

Agamben's engagement with appearance as a theoretical concept and political reality as distinct from inquiries into representations of underlying 'real' objects is part of a long debate in philosophy with the aim to know, with certainty, what exists. Russell claims that this distinction between appearance and reality, that is, between what things seems to be and what they are, is one of the most problematic distinctions in philosophy. Russell's understanding of appearance, one which Agamben challenges, creates a binary between the reality of things - which is comprehended through logic and reason - and the perception of things - which is gathered through the senses. This binary not only privileges reason and mind over senses and body, but also categorizes appearance as representative instead of performative. Agamben's criticism of this appearance-reality binary, guided by his project of causing appearance itself to appear, is grounded in questioning assumptions of an underlying reality from which sense-data and appearance emanate. While Russell's distinction between appearance and reality both presupposes and reinforces the hierarchy between the authoritative subject - that who makes the distinction between the appearance of a thing and the reality of a thing - and the thing itself - that underlying, real, object - Agamben rejects this hierarchical distinction in two ways. First, he does not place either one's self or other subjects as objects to be known. Second, he does not understand appearance as emanating from such objects. These assertions thereby challenge the hierarchy between representing subject and represented object: appearance itself cannot appear - that is, appearance cannot be freed from being a mere effect of the real - if the primacy of the viewing subject is posited as that which receives and processes the sense-data emanating from an object. This not only posits a causal structure between the real and appearance, it also reifies the power imbalance between subject and object. Complicating this power imbalance which, in performance, is between the viewing audience and the performers, is essential for freeing performance from its status as mimetic representation and for returning appearance to itself.

Both Koosil-ja and Brown's performances contribute to this task of making appearance itself appear and realize that acknowledging the audience's gaze merely reinforces the audience's status as authoritative subject and their own status as objects, representing an underlying real. Koosil-ja and Brown's denial of the audience's gaze is essential because, while they acknowledge the audience's tendency to seize hold of their images and make them signify something about an underlying reality, they do not meet the audience's gaze and thereby do not, as Agamben writes, '*show themselves to be simulating*' (2000: 94, emphasis original). Even though, according to Agamben, in the precise moment when 'the actors ... knowingly challenge the voyeur's gaze and force him to look them in the eyes ... the one who looks is confronted with something that concerns unequivocally the ... very structure of truth' (2000: 94), it must be noted that he is describing the situation of actors in front of a camera, not live bodies in a stage-space. The difference is crucial and will lead to a slight revision of Agamben's claims: merely acknowledging that one's appearance is artificial, even if this lends some sort of authenticity to one's image, reifies the binary between underlying material, physiological body and a false, artificial appearance emanating from this body. However, as will be discussed below, in order for appearance itself to appear, it must not acknowledge that it is the projection of some underlying, material body, but instead engage with this material body and dwell in the space of struggle between appearance and the moving, material body.

In film, the close-up is the quintessential

seizure of the actor's image: the camera - the literal mechanism of the gaze - being ignored by the actor at such an intimate range amplifies the artifice of this appearance, and a rupture in this artifice, by the actor acknowledging the gaze, lends authenticity. As Agamben notes, 'the image appears more convincing if it shows openly its own artifice' (2000: 94). And this is accurate: if the actor in a film suddenly addresses the audience, with a monologue or a knowing glance, the actor underlying the character on-screen is revealed to the audience and the actor herself is lent authenticity. However, while a screen-actor who acknowledges that she is acting by meeting the gaze of the audience lends authenticity to her image, dancing bodies in a stage-space have a different relationship with the audience's gaze. In traditional ballet, for example, acknowledging the audience, by say, running from the wings and pausing at the front of the stage with a smile and nod before a solo, might be understood as lending authenticity to the image - the story of 'story ballets' exposed as an excuse for variations of technical virtuosity danced explicitly for the consumptive enjoyment of the audience and the dancers, for example. But these examples of authenticity are not ones of human beings shining in their appearance, communicating only pure communicability; no, these examples of ballet and screen film revealing their artifice through acknowledging the gaze of the audience and gaining authenticity through such acknowledgment merely reifies the hierarchical dominance of the viewing subject over the viewed object. In other words, even though the image itself - on the camera or in the traditional stage display of virtuosity - may gain authenticity by acknowledging its construction for an audience, the subjects of these images are still held captive by the seizure of their image by the audience.

Various mechanisms, especially in postmodern dance performances, have been developed in order to undermine this objectifying force without participating in an acknowledgement of the audience's gaze in order to reveal the performance's artifice, including abandoning the traditional stage-space in which the audience is an anonymous voyeur in the vast darkness and introducing performance practices to more informal or site-specific locations in which the audience and performers share the stage and audience space. Others have attempted to break the barrier between the audience and the performer in the traditional stage-space by bringing the performers into the audience space and vice versa. Brown and Koosil-ja also challenge the objectifying function of the traditional stage-space, the former by keeping the house lights up for the solo and by turning her back on the audience, the latter by breaking up the stage-space into two and introducing live-feed cameras so the audience is faced with overlapping performers - one virtual, one real - in constantly changing combinations. But such challenges to objectification do not, on their own, counteract the desire to take hold of appearances. They are, therefore, engaging in the process of returning appearance to itself: their acts of challenge against the audience's objectification are not attempts to reclaim their appearance for their own possession but are instead performative processes in which their appearance is made manifest as material reality and their subjectivities thereby constituted. Brown and Koosil-ja, in different ways, manifest their material subjectivities in and through their appearances in performance. Therefore, Brown and Koosil-ja's complication of and challenge to the objectifying impulse of the audience's gaze on the performing body and denying the subsequent taking possession of this body-turned-object is a political act, as well as an aesthetic challenge.

Live performance and moving, dancing bodies especially, therefore, pose a challenge to understanding appearance as a merely significatory mechanism by complicating the neat distinction between 'what is' as 'real' and 'what appears' as mere emanation of this real: is the 'reality' of performance the performers qua performers onstage or the scenario and characters portrayed through the performance?

Is the performance only 'real' if the dancers' projected appearance is captured and comprehended by the viewing audience? Is the 'appearance' of performance the perception of a third-order portrayal of reality? Some might argue that the choreography (or script in the case of the screen actor) is the 'real' object of performance and that the embodiment and enactment of this 'real' script in front of an audience is imitation emanating from an underlying real. However, *Dance Without Bodies* challenges this supposition: the images on the small television screens, which Koosil-ja and Guerrero takes as cues for their movements, are presented in a different, random order from performance to performance. So, the underlying 'real' of the choreography is unstable and constantly changing and does not exist apart from the performance itself.

Still others might argue that the underlying 'reality' of the performance is the technical and material attributes of performance - set, stage, costumes, lighting, sound etc. - which adorn and transform actors into various characters. However, relegating the 'real' to the underlying material attributes of the performance and having such technical attributes transform the actors into images in the stage-space participates in this binary between real objects and seeming projections of the real as well as participating in binary between the underlying, material bodies of the dancers and their projected images to be consumed by the audience. Brown's piece however, in challenging this assertion, begins with the house lights at full power, the curtain raised and an eerie, improvised, experimental piece of music being played thereby extending the acoustic attributes of the performance into the 'real' space of the audience and complicating the boundary between technical attributes and an image created for audience consumption.

Brown and Koosil-ja, in and through their rejection and fragmentation of the audience's gaze, their resistance to merely enacting a pre-determined, choreographic 'real' and their rejection of traditional boundaries between the 'real' space of the audience and the 'projected' stage-space, complicate the binaristic distinctions between an underlying reality and the appearance of this reality. Their appearances exist, not as representations of an underlying reality, but as nothing more and nothing other than appearing, just as 'pure communicability', according to Agamben, does not convey an underlying meaning and is, instead, nothing more and nothing less than communication itself. Thus freed from the hierarchical divide between object and subject, appearance and underlying reality and 'what is' and 'what seems to be', appearance is likewise freed from the task of representation and, instead of being in the business of aesthetic projections, representations or mimeses of reality, is faced with Agamben's task of making manifest appearance itself. The question, therefore, is no longer, 'what does this image of dancing bodies signify, mean, represent' but 'what does this appearance do, what are its affective, material, political effects'. And finally, the question is also, moving into Agamben's work on the open, how do appearance and the physiological, material, moving body exist together, not in a hierarchical distinction (i.e. the material body as more real than its projected appearance) nor in a causal relationship (i.e. appearance emanating from the material body), but in a point of momentary repose in a dialectical struggle, of both creating this open space between appearance and materiality and dwelling in this open.

If the question of appearance in performance is therefore not what appearance means, signifies and represents, but what appearance does, how it functions and what it brings into being, how does this affect notions of subjecthood and agency especially in relation to the performing body and the moving body in general? As was discussed above, for Agamben this relationship between appearance and human subjects is located in the space of 'pure communicability': humans 'seize hold' of their appearance though language as a significatory mechanism that can point to one's appearance and identity it. This, according to

Agamben, is the role of politics. He writes: 'What remains hidden from [human beings] ... is not something behind their appearance, but rather appearing itself ... The task of politics is to return appearance itself to appearance, to cause appearance itself to appear' (2000: 95). However, what is the relationship between moving bodies' materiality and their appearance and what are the political stakes and implications for performance? How does understanding appearance as dialectically struggling with materiality, instead of as a symbol of representation, challenge binaristic separations of appearance and reality, represented object and representing subject, and enable appearance itself to appear without denying the political importance of bodily materiality? To explore these questions, I would like to formulate a working understanding of, what I will provisionally call, 'the materiality of appearance' as that which lies between materiality and appearance, the politics at work in the body's appearance in and through performance, and how appearance is at work in the process of subjectification.

In order to move away from the hierarchical distinctions between reality and appearance and subject and object and towards a conception of appearance both freed from this binary and simultaneously engaging the body's materiality and affective potential in and through the process of performance, let us first understand the political as well as theoretical imperative to free appearance from its role as representing or signifying an underlying reality. Representation, far from innocently reflecting an always-already established reality, actively constructs that which is being portrayed and, in this process, takes possession or seizes hold of this object of representation. This representation, functioning in the visual medium the same way Agamben describes significatory language, as opposed to pure communicability, as seizing hold of one's image, is a political act with Agamben asserting 'State power today ... is founded above all on the control of appearance' (2000: 95). Feminist critiques of representation, both in the postmodern configuration of representation as all that exists and in the poststructuralist attack of representation as standing in, marking or a token of something other than pure sensation (Colebrook 2000: 48), have asserted that in performance and performance art, in which the object of representation is a living, moving, breathing, material human subject, instances of representation - as opposed to appearance itself - are violent acts of dominance and possession.

Agamben's project of returning appearance to itself - that is, looking to appearance not as resembling or representing the truth, but as truth itself (2000: 95) - challenges this hierarchy between appearance and the real as well as complicates the role of representation. Human beings, according to Agamben, separate the image from the object of the image by taking hold of and controlling appearance. Art is a prime example of this: by portraying an object by re-presenting its appearance via an image, art takes possession of this appearance. Agamben charges 'Politicians, the media establishment, and the advertising industry' (2000: 95) with engaging in such violent acts of separation and control and further claims, against Russell, that such a disjunction between an object and its appearance is misguided. He writes that the true, the real, the proper are 'not something of which we can take possession', nor do they have 'any object other than appearance' (2000: 97). In other words, appearance is not the sensation or perception of the projected real and truth is not obtained through taking possession of the real object through physical holding, mental possession via logic, or by capturing appearance through representation; instead, the true and the real, for Agamben, is appearance itself.

Unlike Agamben, whose aim is to return appearance to itself in order to free appearance from political control so that humans may shine in the opening of pure communicability, Russell's inquiry into the 'independent existence of objects' is motivated by the need to 'be sure of the independent existence of other people's bodies, and therefore...of other people's minds' (1912: 17). Russell's concern, in other words, is not the nature of appearance itself nor the existence of human subjects per se but the problem of accessing reality and possessing knowledge of the independent existence of objects when faced with the appearance of objects. If the appearance is merely 'a sign of some "reality" behind [it]', Russell asks, do we have 'any means of knowing whether there is any reality at all?' (1912: 16). The object, in this framework, is known the way a table is known - through outward characteristics, such as colour, texture, etc. or, for human subjects, through outward identity markers such as race, gender, etc. The reality of another human subject is manifested through outward appearance that reflects or is a sign of the real existence of an embodied mind. Furthermore, and even more pertinent for this discussion of a materiality of appearance realized in and through moving bodies, is the assumption of a subjectivity underlying Russell's inquiry: the human subject is an unchanging, bounded, unified entity that defines itself in and through its rational capacities. Such a theory of subjectivity participates in what Agamben calls the 'metaphysical myth of conjunction'. He writes that 'man has always been thought of as the articulation and conjunction of a body and a soul, of a living and a *logos*, of a natural (or animal) element and a supernatural or social or divine element' (2004: 16).

Agamben challenges Russell's assumptions and instead asserts that this metaphysical myth of man's conjoined nature is both created and sustained by what in *The Open* he identifies as the anthropological machine of humanism and in *Means Without Ends* can be described as humans' seizing hold of their own appearance through significatory mechanisms of language or representation. Determining the place where these two poles of human existence - man and animal, body and appearance - meet in dialectical struggle, though not conjoined into a synthetic unity, is what Agamben calls the open writing that 'all living beings are in the open: they manifest themselves and shine in their appearance' (2000: 91). Appearance itself, thereby, is made manifest in this space of opposition, a space in which dialectical convergence never - for Agamben and contra Heidegger - occurs and there is instead constant dialectical struggle, suspended and shining in momentary repose, between 'what appears' and 'what is', between objectifying subject and objectified object, between appearance and materiality.

Even though Agamben acknowledges, with Russell, that human beings 'separate images from things and give them a name precisely because they want to recognize themselves, that is, they want to take possession of their own very appearance' (2000: 93), this separation between underlying reality and image is, for Russell, essential to access objective truth. This separation for Agamben, by contrast, 'transform[s] the open into ... the battlefield of a political ... whose object is truth [and] goes by the name of History' (ibid). For Agamben, in other words, 'what human beings truly are is nothing other than this dissimulation and this disquietude within the appearance' (2000: 94), what Benjamin calls the dialectic at a standstill, the 'saved night': like the struggle between the audience yearning to see Brown's face in every movement and Brown's insistence never to yield to this desire and like the audience confusedly chasing the fragmented images of Koosil-ja and Guerrero and trying desperately to catch their frantically shifting gazes, it is the struggle between image and self, between appearance and underlying reality, between man and animal that, when suspended in an unresolved opposition, is truth. It is also in this dialectical standstill, this 'saved night', in which man's nature exists in relation to, though not convergent with, its animal nature and also in which appearance itself simultaneously dwells with, though not in hierarchical opposition to, the material bodies of this appearance.

Therefore, it is in this space of disquietude in appearance, I would like to claim, that the possibility for a materiality of appearance - a moment of repose in the dialectical struggle between appearance and underlying materiality - to crystallize exists. Both of these dance pieces work with the body and the materiality of appearance in different ways, both in their choreography and their uses of the stage-space(s), but both ultimately create and inhabit that opened space of momentary repose in the 'saved night' between two contrasting poles of appearance and materiality, audience and image, objects and subject. And finally, just as this point of suspended dialectical struggle of man's animal and divine natures arrests the anthropological machine of humanism which uphold the metaphysical myth of convergence, so does this moment of disquietude within appearance itself arrest the objectifying and dominating gaze of the audience which seizes hold of dancing bodies' images as significations or representations of an underlying real.

By engaging with this idea of a 'materiality of appearance', I therefore mean to emphasize that the task of causing appearance itself to appear is not, and cannot be, an act of idealism: appearance must appear not as a mere inversion of the binary between appearance and representation nor as a rejection or transcendence of the materiality of human bodies which risks being lost in Agamben's dictum to 'not remain the subjects of your properties or faculties, do not stay beneath them: rather go with them, in them, beyond them' (2000: 100). Appearance must appear, instead, through a contested process of constant struggle between their animal and human natures, between consumed object and authoritative viewer, between what appears and what is. Brown and Koosil-ja's performances enact this process of struggle, moving within a contested stage-space, moving against the force of the audience's objectifying gaze, and moving constantly as simultaneously subject (agent of their own movements) and object (observed dancing body), underlying material body and image, and recognizable human subject and unrecognizable material entity. About three minutes into watching Brown's performance, her body becomes almost inhuman: without her face, she is literally unrecognizable and her moving body is freed from representing an aging, female, performing body to pure moving materiality, to pure communicability. Koosil-ja and Guerrero are likewise unrecognizable, partly because their faces are obscured with goggles and skiing hats and partly because they keep switching sides of the stage-space and it is difficult to tell one from the other between their unmediated bodies and the mediated bodies on the live-feed screen.

Therefore, in order to make their appearance real, not a symbol subject to political control nor as appearance transcendent of material burdens, but instead as material appearance itself, Brown and Koosil-ja achieve the dual task of causing appearance itself to appear while simultaneously accepting the material reality of their moving bodies, but real in and through this appearance itself. In accomplishing this task, the appearance of the body in performance is not understood as representing something other than itself, it is not a symbolic 'stand-in' for reality, nor is it functioning merely on the level of mimetic imagery; instead, appearance is itself a material reality. Therefore, at stake in acknowledging the materiality of appearance is not only freeing performance from the class of false, mimetic, illusionary representations of reality but returning appearance to itself, realizing that appearance and material bodies, especially in performance, exist in a dialectical struggle, the process of which actually constitutes relationally defined subjectivities and affective interactions in and among dancers and observers instead of merely representing always-already constituted subjects, and finally, acknowledging that appearance itself belongs, not to the realm of 'what seems to be', but to the realm of 'what is'.

Let's try again. A performer appears onstage, her back turned towards the audience, and begins to dance. She moves constantly, her body past her youth, her shoulder blades and spine highlighted by the overhead stage lights. The mechanics of her sweeping movements are articulated by her skeletal back. She never turns to face the audience for the duration of the solo - they know her, not through her face or expression, but through her appearance onstage. When she finally faces the audience to bow, her face seems foreign and unfamiliar, as if being greeted by a stranger. By turning her back to the audience and never meeting its gaze with her own, Brown denies the audience the possibility to turn her into a significatory mechanism - that is, a 'mere' representation of underlying reality - while simultaneously revealing through her performance, not something underlying her image, but appearance itself. But this time the audience is not frustrated by her hiding. This time, they welcome her moving body, the appearance of her moving body, as a processual, material becoming. She stands in for nothing but her own self in the process of revealing not an essential, pre-formed self, but revealing the process of her own coming into being. In and through her performance she both holds her own image and manifests herself through her appearance. This time around, the audience does not observe this process as sovereign pre-formed subjects for, in this version of the story, neither performer nor audience exist prior to the process of performing, the process of observing. Instead, the audience themselves are cultivated and made manifest through dwelling in the open afforded by Brown's performance. They interact with her shining appearance in an affective relationship in which neither viewed object nor viewing subject are already formed or dominating or representing or represented. They interact with her appearance as sensation not as representation or signification. They experience her moving body, not as a representation of her subjectivity, but, as Agamben writes, as 'a passion' - an affective interaction - 'of revelation' and of appearance (2000: 92). It is this act of appearance appearing, not as a representation or signification of some underlying reality because, as Agamben writes, 'human beings neither are nor have to be any essence, any nature, or any specific destiny', but as appearance itself that is, for Agamben, 'truth' (2000: 95).

Let's try again. Fast forward twelve years and enter into a stage-space split into two with bleacher-seating on each side, back-to-back and live-feed cameras portraying the activities occurring on the opposite side on a large screen. Momentarily confused as to whose mediated-depiction is being viewed on the large screen, the audience is faced with a reflection of another audience's appearance and another audience's gaze. Koosil-ja's transformed stage-space frustrates the audience's role as unobserved, authoritative viewer and the images of choreographer Koosil-ja and fellow dancer Melissa Guerrero's moving bodies, mediated by the live-feed cameras, cannot be anchored or seized in the process of performance. Koosil-ja and Guerrero, instead of overtly denying the audience's gaze as Brown does, complicate the audience's gaze by constantly changing sides of the stage-space and by frantically shifting their own gaze between five clusters of television sets set up on each side of the stage-space, almost too busy to be bothered with the onlookers trying desperately to catch their eye. This time around, however, the audience is not trying to figure out their own subject positions within the complicated and virtually mediated stage-space nor are they frantically searching for an un-mediated view of both dancers to grasp hold of their gaze and thereby access their pre-formed selves through their representations in performance. Instead, the audience becomes part of an extended nervous system of the performance space, the technological elements and the performers. They all dwell together in a constantly transforming space with bodies, images, gazes all struggling against the others to weave together into some sort of openness, struggling to momentarily repose into some coherent appearance incorporating all materialities, all images, all gazes. The question is not what is represented on the large screen - one side of the audience is constituted only in reciprocal recognition of the other - nor what is represented by the movements of the performers. The question is how to recognize the performers seizing hold of their manifest beings in and through this mediating technology and then recognizing the audience itself as part of this shining in an interconnected environment of appearance.

With Brown and Koosil-ja's recognition of their own images and dwelling in the space of the open, appearance appears unfettered by connections to an underlying reality or to possession by an authoritative viewing subject. Appearance appears as sensation and affect and this open subject - this subject that manifests itself in its appearance - produces sensation when it acts on and between the bodies of performer and audience. In two very different ways - Brown by denying the audience's gaze and Koosil-ja by fragmenting this gaze - these dance pieces deny the objectifying gaze of the audience and create an affective relationship between performers and audience that is based in sensation, in appearance-as-appearance instead of appearance-as-signifier. When appearance itself appears and is freed from representation and its connection to an underlying reality, its own materiality is realized: instead of abstract signs pointing to an underlying materiality, the sensations of appearance struggle in dialectical movement between and among the material bodies of the dancers and audience members and the task of causing appearance to appear is a task of recognizing the materiality and reality of sensations. The process of seeing - the interaction between light rays and the receptors in your physiological eyes and then passing through the nervous system into the brain - is a material process. In these performances, similar to the systems of the physiological experiencing sensations via appearance, the bodies of both the performers and audience members are opened 'to connections that presuppose an entire assemble, circuits, conjunctions, levels and thresholds' (Deleuze and Guattari 1987: 160). Through such interconnectedness, 'logical consistency [is given] to the in-between', to those occurrences between performer(s) and audience, and the subjectivities of the performer(s) and audience are cultivated, not through an essential separation between subject and object or appearance and reality, but through a 'being of ... relation[s]' (Massumi 2002: 70). And finally, 'When sensation is linked to the body in this way' - when sensations are part of an extended process of experiencing appearance and affective interactions and when the subjectivities of the performers and audience members are cultivated through their relational instead of hierarchical positioning - appearance, and the sensation experienced of this appearance, 'ceases to be representative and', in a completion of Agamben's project, 'becomes real' (Deleuze 2002: 40).

REFERENCES

Agamben, Giorgio (2000) *Means Without Ends: Notes on Politics*, trans. Vincenzo Binetti and Cesare Casarino, Minneapolis: University of Minnesota Press.

Agamben, Giorgio (2004) *The Open: Man and Animal*, trans. Kevin Attell, Stanford: Stanford University Press.

Colebrook, Claire (2000) 'Questioning representation', *Substance* 92: 47–67.

Deleuze, Gilles and Felix Guattari (1987) *A Thousand Plateaus: Capitalism and Schizophrenia*, trans. Brian Massumi. Minneapolis and London: University of Minnesota Press.

Deleuze, Gilles (2002) *Francis Bacon: The Logic of Sensation*, trans. Daniel W. Smith, Minneapolis: University of Minnesota Press.

Massumi, Brian (2002) *Parables for the Virtual: Movement, Affect, and Sensation*. Durham and London: Duke University Press.

Russell, Bertrand (1912) *The Problems of Philosophy*, Indianapolis and Cambridge: Hackett Publishing Press.

KINKALERI

Wanted

intuitions
about
the
world
waiting
to
become
an
established
construction

I - Boxed Wonder
II - No Wonder
III - Invisible Wonder
IV - The Last Wonder

Kinkaleri settle down in Galleria Accursio, a volume underneath the surface of the historical centre of the city of Bologna, a ticketless place of apocalyptic construction of the wonder.

Different displays of exposure of practices of the presence in a place always reshaping itself by a flux of performative events, housing distensions and relaxations, soundscapes, items for contemplative states.

Wanted composes itself of its own pulverization, in a utopian way; it develops in four days distinguished by variable contents and durations of expanded times, arranged in a unique long fragment that includes important collaborations and relations with prestigious international artists and critics: Federico Bacci, Romeo Castellucci, Forced Entertainment, Barbara Manzetti, Kinkaleri, Nico Vascellari, Fabio Acca, Joe Kelleher.

Questions, demands, informations, exhibitions, answers, orders, resentments, intentions, simulations, attitudes, instructions, pleas.
Not on the edge this time, rather on one desolate land crowded like a city.
Every act of Wanted refers to one political presence that shows itself for carrying out a duty and solving one work.
A cannibal act, beyond the result.

Wonder is a bomb

Performance Research 13(4), pp.48-55 © Taylor & Francis Ltd 20
DOI: 10.1080/135281609028756

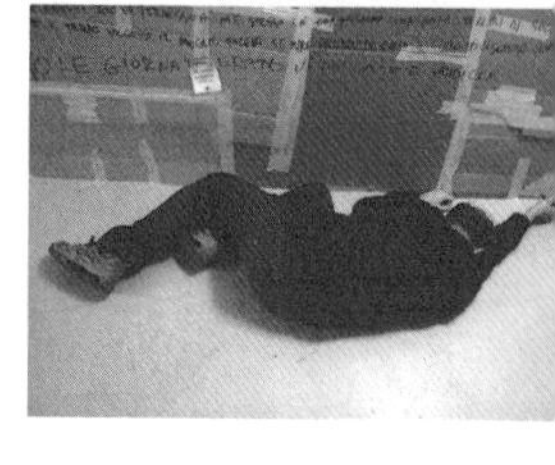
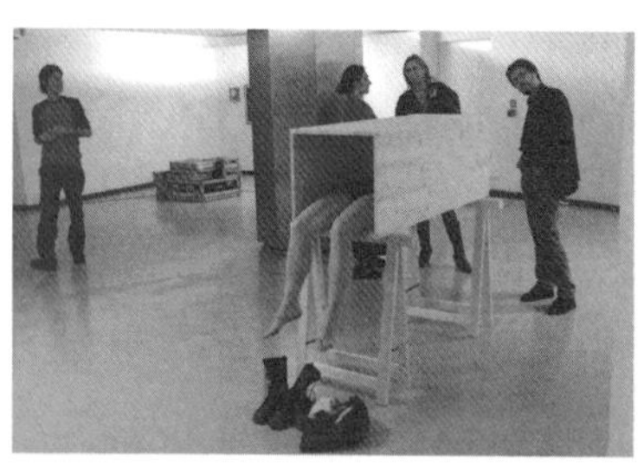

Cosa
facciamo
dopo?

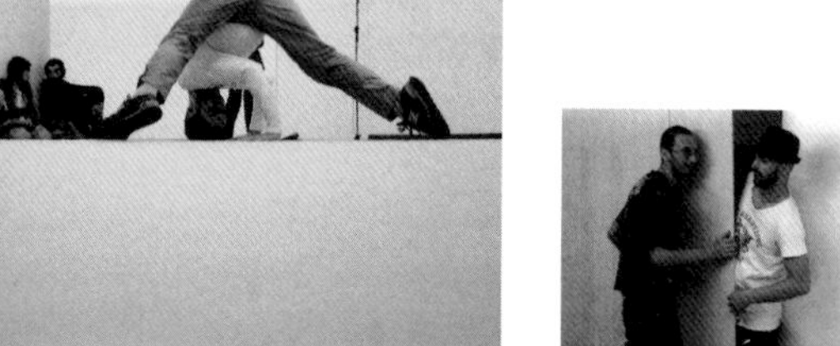

The jungle is important. It has to do with sense, something very close to peoples' hearts. They aren't my words, but come from a dancer friend of mine. After one of his shows C. S. flung himself on me. 'You've got to explain to me right now what this is all about. It's neither dance nor music!' The only thing I could think of was my mother, when she'd say that Hendrix wasn't music but noise. Although the only thing she listened to was Mina. Basically, the sixties passed over her without leaving a mark. After the show I was telling this to my friend, who said to me, astonished, 'So if you see a snake, a jungle, do you ask yourself what it means? It's you looking at it, but it's your body that makes sense of what it's seeing. Because it's the body that responds physically to another body. With amazement, wonder or perhaps fear. We dancers are like snakes in the jungle.' Perfect answer.

Wanted is also an exhibition of snakes, objects, bodies, actions, silences, expectations, to be responded to as it comes.

It is a threshold upon which attention fluctuates, and then freezes stupidly the instant the body believes the moment is right.

It is an investigation of opening, of welcoming, of the anthropological necessity – I'd say – of exposure. The audience, being a stranger, responds 'naturally' to its call.

It is the dead zone of the will, its black hole.

Fabio Acca

extract from 'La zona morta della volontà' published in Today is ok, edited by Xing, Bologna, 2007 [translated by Kinkaleri].

WANTED

Galleria Accursio, Bologna, 23 - 26 April 2007.

A special project by Kinkaleri with Fabio Acca, Federico Bacci, Romeo Castellucci, Forced Entertainment, Joe Kelleher, Kinkaleri, Barbara Manzetti, Nico Vascellari.

An initiative of the Siemens Arts Programme in cooperation with Xing/F.I.S.Co.07.

Kinkaleri was founded in 1995 and has since produced plays, performance pieces, installations, videos, soundtracks and publications, all characterized by mixed language and transverse signs which invite rapid redefinition.

Kinkaleri is based in Prato, Italy, at Spazio K. www.kinkaleri.it

Kinkaleri has been asked to hand over the baton of *Wanted* to the company MK & guests in 2008 (who created Wasted at F.I.S.Co. 08, Bologna) and BAROKTHEGREAT vs guests (who created *Wrestling* at F.I.S.Co. 09, Bologna) proposing them to reinterpret its modality and spirit.

Photos by Gaetano Cammarota, Roberto Serra/Xing, Luca Ghedini, Kinkaleri.

How to Act, How to Spectate (Laughing Matter)

JOE KELLEHER

1. EVERYBODY ALWAYS ACTS

For the critic Fredric Jameson the fundamental 'method' of the author and theatre-maker Bertolt Brecht, which Jameson characterizes at one point in his book on the topic as 'a kind of ethos, or at least a moral training of a specific type', has to do with bringing to appearance the forms of 'productivity' that inhere in human 'activity' (Jameson 2000: 53-4, 177-8). In the theatre that involves a technology of making-appear that focuses on the very activity of bringing to appearance as such, for example in the sort of pedagogic theatre that stages 'the showing of showing, the showing of how you show and demonstrate' (91). The implication here is that what is to be shown - or said - about something is already given, or rather already being produced, in the activity that is being represented: an implication that applies not only to the specialist knowledge-producing activity of a historical figure such as Galileo but also to 'those instances in which the theoretical content of our everyday movements suddenly intrudes upon us and our fellow "actors"' (84). In what I take to be a key passage of his book, Jameson characterizes this uncommon potentiality of the most common activity with the term 'proto-dramatic'. It is a matter, Jameson writes,

> of what people say about what they do - in other words, about the inherent and verbal knowledge their gestures and actions carry with them, and how they explain these to themselves and to others. In that sense, we may also suggest that for Brecht, too, everybody always acts, and that we ceaselessly tell stories to explain ourselves, dramatizing our points in all kinds of ways, the undramatic as well as the ostentatious and self-parodical. It is thus better to shift the vocabulary of reflexivity, and to suggest that all acts are not so much reflexive and self-conscious as they are already proto-dramatic. (83)

Of course, in these 'post-dramatic' times, we might not approach this insight - which is essentially an insight into the sort of activity we have learned to designate with words like 'performative' and 'performativity' - with the same sort of hopes and intentions as Brecht. Jameson himself insists that there is a 'central difference between Brecht (and his whole era) and our own *Zeitgeist*', a difference to be located in a distinction between 'opposition' and 'contradiction', which is to say in the moment in which 'class conflict (Opposition) becomes genuine revolution as such' (79, 82). As for genuine revolution on the theatrical stage, we will continue to look forward to it. That said, there may still be something to say about how the Brechtian acting lesson has transmuted, in our own times, into ways of being on stage (I think immediately, to name some of the more heralded examples, of the work of the New York City Players, or TG Stan or Forced Entertainment or Needcompany or the Wooster Group) in which, if the theatrical method doesn't necessarily engage contradiction so much, still it does involve a presentation (a highly refined presentation in many cases) of the proto-dramatic potentiality of everyday activity. The distinction being that this is a potential that hasn't quite been rendered dramatic yet, hasn't

Performance Research 13(4), pp.56-63 © Taylor & Francis Ltd 2008
DOI: 10.1080/13528160902875630

quite been offered up as image-worthy *yet* - as if to be caught up wholly in the economy of images and dramas, with nothing lacking or nothing left over, were somehow to be estranged (and not in a progressive, Brechtian way) from the productivity of one's own appearance.

It is in relation to this 'not yet' that the term proto-dramatic may serve us better than the term performative, although that is not to suggest for a moment that there is anything less than rhetorical (not even anything less rhetorical than the modernist theatre of Brecht) in any of the practices name-checked above. The theatrical rendering of the proto-dramatic is altogether rhetorical, although it is also worth remarking that a key sign of that rhetoricality on the contemporary stage - as opposed, perhaps, to the supposed political seriousness of the Brechtian dramaturgy - is laughter. That might be the scripted 'Huh huh huh' in a play such as Richard Maxwell's *Joe*, where the laughter functions as a floating symptom - shared between the various *characters* - of a sort of re-distribution of subjectivity across the various parts of the drama; or the seemingly accidental and uncontrollable (but no less rhetorical) laughter that erupts among the *actors* in a performance by TG Stan or Forced Entertainment, functioning like a sort of meta-commentary upon the temporal condition of being on stage and having to take responsibility for what is going on (see Ridout 2007: 130f). In any case the laughter is asking a question of how one can possibly act in such a situation, I mean the sort of situation where the productivity of appearances is at stake: for example the intolerable situation, prevalent enough these days, of 'being' - or being taken for - a living image.[1] Our question here - and it will remain in all sorts of ways a 'political' question - may be the following: how does an image act? And beyond that, what might an image that acts - under whatever sufferance - have to say for itself?

2. LAUGHING HOLE

The three performers are laughing already when we go in. There are barely as many of us as there are of them. The floor is covered with strips of brown corrugated cardboard and there is nowhere obvious to sit, or stand. Maybe that's what they're laughing about. We find spots to place ourselves, away from each other, close to the walls. Over the six hours of the performance at the Toynbee Studios in East London, in an upstairs room that has the look of a former class-room or meeting hall, parquet floored, white walled, with two supporting columns in the centre and windows on three sides of the room letting in the sunlight, we will come and go. Our number will increase and diminish. The three women meanwhile - choreographer La Ribot and two accomplices - remain in the room, doing what they have to do and laughing as they do it, continually laughing. It would appear that we are hardly there for them, or at least what they have to do is not being done 'for' us. Several of us take out notebooks and start scribbling. As well as the three performers there are two male sound technicians, at desks placed prominently in the room, diagonally across from each other. For the first half-hour or so, though, the womens' raw laughter is the only sound going on, that and the sound of sticky tape being torn. The performers' business has to do with those strips of card that make a thick carpet across the floor. Each of the three performers in her own time falls to the floor, picks up a strip of cardboard and strikes a pose, exposing the text that is felt-tipped in capitals on the other side of the card. She then tapes the strip of cardboard against one of the vertical surfaces in the room - walls, windows, doors, pillars. Each text is a two- or three-word phrase, like an image caption, or a speech bubble, or a piece of signage, the title of an event (for example, the title of the present event LAUGHING HOLE, one phrase from among others placed over the event as if such a designation were not, at the same time, a sign of the indistinction of any such designations: animal, vegetable or mineral - under such conditions who is to say). As more of these strips are taped to the walls it becomes clear quickly that the words that are spliced together to make up these phrases are chopped out of a lexicon of

[1] A territory that was, of course, mined exhaustively by several twentieth-century theatrical authors, such as Samuel Beckett

recurring terms: ANONYMOUS, FALL, OVER 40S, DUCK, OPERATION, BRUTAL, MUM, ALIEN, DETENTION, ILLEGALLY, DEATH, MISSING, LOST, SPECTATOR, SOFT. Hence, FOR DETENTION, CLEAN FALL, STILL IN DETENTION, MY FLIGHT, FEED THE DUCKS, BRUTAL HALL. Although the phrases don't quite function as names or statements - they seem more exhausted or more impatient than that, epithets slapped against the surfaces of a recalcitrant reality in the hope that something might stick, or in fear that it might stick too well - the accumulative effect of the words' appearance on the walls is like a gathering up of the lexical waste of our contemporary state of exception.[2] There appears to be no way anymore of sifting one thing from another, no way of 'treating' the languages we share with each other that might save them - and us - from the lexical infections of our times. Hence, perhaps, the seemingly endless task of scooping up all these spent phrases and pressing them onto the walls, doors and windows like a sort of dirty protest.

[2] Philosopher Giorgio Agamben analyses the state of exception as 'the original structure in which law encompasses living beings by means of its own suspension', as for instance in the '"military order" issued by the president of the United States on November 13, 2001, which authorized the "indefinite detention" and trial by "military commission"... of noncitizens suspected of being involved in terrorist activities' (Agamben 2005: 3).

Meanwhile the laughing goes on. All three performers are mic'd up and their giggles, snickers, squeaks and gurgles are played through four large speakers turned into the corners of the room, the voices processed by the engineers into a self-perpetuating ambience of feminine noise, a multiplying vocality that is capable of detaching itself from the people who are making the noise, even as we see the effort it takes, the effort being made right now by these performers, to force this sound to appear. La Ribot has used this trope of continuous laughter in other recent work - specifically the show *40 espontáneos* that was put on in several cities around the world, always in short order with just a few days preparatory workshops, with companies of untrained and largely inexperienced performers recruited in each case through local advertising. There the laughing looked and sounded (on the occasion I saw the work in London) like difficult labour. These amateur performers' ostensible commitment to the task of keeping this laughter going over the hour of that work emphasized - if anything - the theatricality of the gesture, if we might understand by the term 'theatricality' a distancing (emotional and also physical) on the part of spectators and performers from the 'clear and present' production of theatrical side- (we won't say side-splitting) affects.

In *Laughing Hole* there is no less distancing involved, although the laughing here feels less forced, or perhaps better 'acted'. Even when the performers come up close, however, leaning against the wall beside us, taping a strip of cardboard up behind our heads, the continuous laughter pre-empts the notion that we might have any other share in what is going on here than the share that belongs to us as spectators. The laughter serves to remind us that the proximity of the three women in house-coats and flip-flops - their being at times close enough to touch, while extending a seemingly open invitation to touch their laughter with our own, to laugh together over this (whatever 'this' is) - is illusory. Their world is not our world. Their words, although they look so similar, are not our words. Their laughter has nothing to do with us. This is an illusion that might persuade us to add to the question 'How to act?' a matching question 'How to spectate?', if it really is the case that an effect of *acting* can be enough to separate us so radically from other people that even their most direct statements, their most explicit gestures, become indeterminate for us.

Maybe it's the way the acting is done. When the performers fall, they don't fall as if pushed. They fall as if drunk. They walk, then stop, and then sink into a collapse. What are they drunk on? Is it the same for each? They work independently of each other, each of them at any moment in a different place in the room, falling, posing, taping up a sign or just lying still, doing nothing. It's a summer evening, and it won't get dark outside until late, although the room is already darker a couple of hours in than when they started. Several of the cardboard strips are starting to block out the windows. All of the strips, as they are taped up, are pressed flush onto the walls, whether that means folded around a corner or a sill or a door handle or wedged into an alcove or indentation. They are papering the walls with phraseology. It becomes like a visual representation of crowd speech, bits and pieces of exclamation overheard, mis-heard, cut into each other: 'meaningfulness', aboutness, intentionality remaining attached to these phrases like a tonal effect, like a ghost of utterance, even as the words appear abandoned by the world (the world has already moved on) upon which reference would have depended.

So what. For the women it is a job of work. In their housecoats they are dressed for labour. (The flip-flops I read now as some absurd regulatory element of work uniform, or penal uniform). Sometimes they speak the phrases on the signs they are holding, but not as if these utterances are their own. Each sign is held up with the same dedication to the gesture of holding up, the same indifference. There is no 'showing' the signs to the spectators in the room. They are held up, they are visible; to whom they are visible is neither here nor there, just as it is neither here nor there who these utterances might belong to, although along with the ghost of utterance there are scraps of identification: GAZA, TERROR, ME, YOUR, MY. Everything is legible. Although the signs cannot be taped any higher up on the walls than the performers are capable of reaching, no sign is taped over any other sign if possible, and every inch of accessible space is used. This is legibility, though, abandoned to mere visibility. Meanwhile the room is transformed into a vertical tunnel of phrases, a sort of moulded word-cave sealed in from the world outside, a padded cell of exclamation, a laughing hole indeed. It's a place where - or this is what I was thinking as I left the performance maybe an hour before the end, leaving them to it, having seen where it was going and how it was likely to go on - language is glimpsed in a state of abandonment, where words no longer do what they promise to do, where it is left to the ones who live there, scavengers of sense, to do things with the words. And they live it like adepts, who have got past the point of choosing what to do or how to do it. The acting, if that is what they were doing, has become like a lived relation, as the room is sealed in and reference, as it were, is sealed out. What presses from outside is a great violence. These words on the walls are its evidence, its remains, whatever has leaked in here among us, to be met by a laughing resistance that resists by taking what is given while night descends outside, taping up the words over doors, walls and windows so that night will not descend in here, so that there will still be words, however tainted, with which we can do things, a doing that may amount to no more than preserving a space of speech and action as a place of entertainment, somewhere something is done and something is said; done and said by these abandoned women,[3] 'acting' with abandon, doing what they can with the abandoned gestures, with all of these abandoned words.

3. THE PROMISING ANIMAL

Over the six-hours' labour of La Ribot and her co-performers, a relation is proposed between action, image and language. The insistent force of duration, repetition and accumulation is enough to persuade us that a relation of some sort is at stake, even if it turns out to be the sort of dramaturgically disjunctive 'relation of non-relation' that adheres to the essentially presentational rhetoric of this work (in spite of its distended duration and its 'environmental' *mise-en-scène*). We can anyway say this much: the bits of language that get displayed and put up on the walls require the performers' actions in order to be so displayed, and this displaying is accompanied by the performers' vocal exertions. We can say also that while the meaning - or the promise of meaning - in these deliberate falls and poses, or in the temporally and spatially dilated image of the room itself, or in these sound-bites stuck to the walls, or in the laughter that goes on and on, is not straightforwardly available in the form of a statement, still the performance has about it a gestural force that has to do with our sense of what is and is not promised or delivered in this carefully scattered choreography of utterance, movement and appearance.

3 In Ramsay Burt's reading of La Ribot's 'laughing' works, they are '"destroyed men" (and women)... people undone by the disaster of modernity' rather than abandoned. Burt's essay (2008) came to my attention just as the present article was being completed and is highly recommended, developing as it does - and with considerably more economy than the present attempt - a sophisticated argument about this work as a provocatively resistant form of performed 'passivity'. I am grateful to Charlie Fox for putting me on to Burt's article.

Sometimes, of course, the promise is (seemingly) explicit, and the delivery pretty direct. At one point a sign is held up bearing the phrase THIS IS GUANTANAMO, to which my immediate response is 'no it's not'. This place in London is *not* that place in Cuba. Nor is it, at whatever level of abstraction, a representation of that place. I won't, or don't want to, accept that. That way is far too direct, and to that extent profoundly un-seductive. Anyway, to be 'taking on' the violence of that situation in this way would signal, somehow, too easy a renunciation of the violence that is also going on right here this evening, in this 'theatre' where words and gestures are cut and broken, shaken through a sieve and pressed upon the event like sticking plasters to cover up what appears more and more as the evening goes on to be a (collectively) self-inflected wound. I would even suggest that the phrase ('This is Guantanamo') does not mean what it appears to say, that saying of that sort isn't what goes on here; that the sign cannot *do* that job of stating and meaning, not here, not this evening, not among those phoney drunken falls, not with all that spluttering laughter going on. A certain violence, to which these falls and fractured phrases bear witness, has seen to that. Rather, as the gap opens up between one place on earth and another, between the place where this June evening I am attending a performance in the Artsadmin summer season and the infamous detention centre in the Caribbean where, as Agamben puts it, 'bare life reaches its maximum indeterminacy' (Agamben 2005: 4) and which, more than any other place on earth, if only through the familiarity of its name, 'stands for' the contemporary state of exception, alongside the collapsed distance between this place and that (the collapse is verbal, of course, although it also involves a bending of the knees ...) another sort of gap opens up like a trench, another sort of laughing hole perhaps.

This would be the divide that separates words and actions that are given place in the theatre, i.e., in the space of representation, from the proto-dramatic words and actions that take place in the represented world, a line that is conventionally marked by a row of lights or a sheet of cloth or a raised platform or a designated distance between the knees of the spectators and the lip of whatever fiction is being played out in front of us for our entertainment. Tonight this line is much more indistinct than that, given the way that here we may sit - or stand - *wherever we like*, and given the way too that the performers can come close enough for you to feel that closeness as a promise made not so much to the interpretive reason that you imagine you share with everyone else in the room but to the feeling body that you believe you share with no-one else at all. Close enough, in other words, for the laughter to be an exchange of breath. Even in the eroticized zone of indistinction that the performance sets up, however, we may register a trace of theatrical convention - a trace as it were of historical stuff lodged like an indigestible stone in the throat of the laughing hole - that is sufficient to remind us that an act in the theatre and an act in the world are not necessarily of the same order although they may involve the same gestures, the same costumes, the same sort of words. We may recognize that theatrical violence and political violence are not contiguous with each other although they may well be concerned with the same thing; and that any apparent indistinction between the languages of theatre and politics may itself be an outcome of the same, ubiquitous state of exception whose modern history is contiguous with the history of 'political' theatre and performance, and against which so much of this work has set its sights.

What after all are those three performers laughing *at* or *about*? If we can suggest that they appear to be laughing at what they are doing, or at the very least their laughter is inextricably involved in what they are doing, then we might offer that what is funny about their 'act' would appear to have something to do with a simultaneous excess (such a long performance, so many signs taped to the walls) and at the same time redundancy, or falling short, of gesture and utterance. A falling short - or over-shooting - of utterance (or say an *image* of gesture) that brings bodies along with it, that needs the labour of bodies in order to appear. And what bodies they are: collapsing, falling, sweating bodies, bodies with flushed faces and matted hair, their flip-flops flipping, housecoats riding up and knickers showing, doubled up in laughter, drunk it would seem, practically wiped out, with the sheer, exhausting funniness of it all.

And falling is funny, as Shoshana Felman would remind us, being 'closely tied to the performative, since falling is an *act*: *the* act, indeed, so far as it is a failure - the very prototype of the acte *manqué*' (Felman 2003: 85). Felman's words are from her 1980 book (translated and re-issued in English in 2003) *The Scandal of the Speaking Body: Don Juan with J. L. Austin, or Seduction in Two Languages*, where she reads Austin's elaboration of a theory of performative language alongside a series of reflections on theatrical, or quasi-theatrical, seduction, where seduction is wrapped up in the business of failed - or betrayed - promises, exemplified by the dramatic (and operatic) antics of the mythical arch-seducer Don Juan.[4]

Don Juan, in the Molière and Mozart / da Ponte incarnations, typically, and repeatedly, promises marriage, and in doing so touches on one of the key examples of performative utterance, the 'I do' of the marriage vow. The problem of the vow, however, is that such an utterance is already a promise to bring a body along with it, and although there is no spoken utterance without a body to support it (literally, to say the words), the promise is also bound to be broken to the extent that this - or any other body - as the site of 'unconscious aims' and as the signifier of what Judith Butler calls the 'unintentional', is always bound to mean more (or less) than what it is capable of saying. Promises, then, tend (either through an excess of follow-through or a falling short in the delivery) to fail: the body can't keep up what the words lay down, at least not interminably; but in that falling short or - as Felman characterizes it - the 'misfire' of the performative, something else happens, '*something else is done*' (57).

What happens, or rather what is *done* in this misfire, is what Felman calls the 'act', specifically the act of the 'speaking body'. This act functions altogether 'theatrically' (although it is not only in the theatre that such a function will be recognized or felt) in that it involves a perceptible 'force' (the word is Felman's adopted from Austin [8]), something that is felt upon the senses, something remarkable, something that might evoke a laugh or some other form of ejaculation, that adheres to but is not contained or accounted for by the ostensible rhetorical 'aim' of this or that speech, or this or that appearance, or this or that movement or gesture. In the face of such a force (which may be an effect of strength or weakness), we can't help but take our pleasure and be seduced, although it may be we don't know what is going on. This knowledge deficit is, for Felman, the seductive scandal of the speaking body. 'The speaking body is scandalous to the extent that its performance is, necessarily, either *tragic* or *comic*' (67). She adds: 'The scandal consists in the fact that the act cannot *know what it is doing*, that the act (of language) subverts both consciousness and knowledge (of language).' In so many words, I can't say what I'm doing when I say 'I do', although the saying does something, 'the *speaking* never fails to do'.

We are back, immediately it would appear, with the indistinct seductions of the actor, where again there will be all sorts of falling down and disturbing laughter, a mass of slippery words under our feet that have to be turned over and dealt with, and also - or did we see this coming? - another sound underneath the laughter, that of a deadly grinding, a mechanical, hollow sound, like an invitation to a dinner party in Hell spoken through a mouth of stone. Here, we will take this as the depthless sound the city makes at night beyond the blacked out windows of the Toynbee Studios; or else the whirring of the backstage engines, the churning of the gears of the anthropological machine,[5] that bring to us a world of appearances - and human appearances at that.

4 I am grateful to Laura Bruton for putting me on to Felman's book.

5 The phrase again is Agamben's and refers to conceptual frameworks according to which what counts, and what doesn't count, as 'human' is decided upon. See Agamben 2004.

Back to Felman: 'What is slippery,' she tells us, 'is thus seduction itself ... what *causes skidding*, in language, is especially pleasure itself, inasmuch as it surreptitiously causes one to slide - without being aware of it - outside the realm of meaning, outside the terrain of knowledge' (85). The theoretical performance, for example the Austinian performance, i.e., what he does with his jokey words about how to do things with words, which tends to be ignored by his philosophical readers as something incidental to the event of thought, is, for Felman, 'poetic'. That is to say, Austin's verbal act - an act that 'is not simply the act of provoking laughter but also that of tripping' - seduces the reader onto slippery ground, where all sorts of (dangerous) pleasures are to be enjoyed. Tripping, Felman suggests, has to do with subversion, intended, in Austin's case, 'to subvert knowledge, to call it into question, to cast doubt upon it' (86). The sign, the symptom of this subversion - this gesture that seems to mean more (or else less) than it shows or says - that is aimed 'above all at shaking the institution of prejudice itself, the institution of beliefs or received ideas' (87), is laughter. Felman: 'If laughter is, literally, a sort of explosion of the speaking body, the act of exploding - with laughter - becomes an explosive performance in every sense of the word', not the least of which senses involves an explosion of belief in the promises of men, or indeed in any of 'the promise[s] of the promising animal, even if that animal is oneself' (88). It is a theme developed more recently by Alan Read, writing at the end of his book *Theatre, Intimacy and Engagement* about the 'political potential' in the theatre's exhibition of 'the decisive art of the indecisive human, the lesser animal. The parallax of the performer is the promise that they fail to keep, the paradox of the actor is the promise that is beyond them' (Read 2007: 279).

I shall leave aside for now the question to what extent this - or any other - performance gets close to achieving the task proposed by Giorgio Agamben of stopping the anthropological machine 'and break[ing] the violent hold that law lays upon life'; or whether the *act* of a performance such as La Ribot's - in the way it might appear to shake loose over its six hours something of what binds the word-hoard to the worlds in which that hoard is spent - evokes that 'pure violence', that 'violence outside the law', that Walter Benjamin in 1921 imagined as a counter-act to the law-making violence and law-preserving violence that institutes and sustains the state of exception; or that 'pure medium' or 'mediality without ends', which Agamben again in the early twenty-first century conceives as the political task for all of us as we suffer the globalization of the state of exception in our times (Agamben 2005: 62). A performance - this or any other - is too fragile, too contingent an object, to be measured in such terms; although, for what it's worth, I'd be inclined to suggest that the theatrical economy of even such a 'presentational' performance as this is still too mixed, too tainted with reference and intention and convention and desire, for any talk of purity to stick.

We might, though, want to make something of how the mixture of media and registers and levels of meaning at work here lends itself to a 'convulsive' violence (which is to say too a pleasurable 'brutal[ity]') through which meaning and intention can *fall*, like blasting holes in representation and sliding on through (although it may be one will never have blasted one's way right through words, acts and images), to slide out ... *where?* On the other side? On the other side of the image - which has no other side? Here is Felman again:

> Now the passage from one level to another is not harmonious, gradual, continuous; on the contrary, it is - like laughter - convulsive and brutal. Meaning, in fact, can be accompanied by pleasure only on condition that it fall from one level to another. In other words, the passage from one textual plane to another is on the order of a skid or a fall. (2003: 84)

This passage between textual planes 'from one level to another' reminds me of another, from a text already cited in this essay and which takes us back - once more - to a consideration of 'political' theatre in early-twentieth-century Europe. Here is Fredric Jameson again in *Brecht and Method*, commenting upon Brecht's notes for the singers in *The Threepenny Opera*, attempting to describe the judicious effects of discontinuity produced by the way the actors distance themselves from their action as they shift from one verbal register to the next:

> It is a process for which I am tempted to revive that much over- and misused term the 'sublime', in a rather different sense from that in which it has been normally appropriated: here, some immense liberation from context as such, as we pass *from one register to another* - to be fundamentally distinguished from the traditional notion of the liberation within the sublime by way of its cancellation of internal limits, by either immense size or immense power. (Jameson 2000: 145, my emphasis. See Brecht 1984: 44)

In a footnote, Jameson elaborates:

> In Longinus (and in the tradition that follows), the sublime is always identified with a movement of 'elevation and amplification'; whereas my proposal suggests the simultaneous possibility of a satiric diminution or lowering so dramatic and prodigious as to evoke sublimity in its own right in turn. (162)

In the figurative similarity of these two passages (Felman: 'from one level to another'; Jameson: 'from one register to another') we have, perhaps, a clue as to how we might hold together - if not in the same thought, then in closely adjacent thoughts - the Brechtian political theatre of the early 1930s and what we might take as the political potentiality of the contemporary dance performance of the early twenty-first century. What holds these thoughts alongside each other, I suggest (although it will need another occasion, at least, to unpack this notion with any justice), is an image of praxis, as it happens a laughing praxis in both instances (I will take 'satiric diminution' as implying at least the possibility of a chuckle or two), lodged, oddly enough, in the throat of theory (we are citing Felman and Jameson after all), but threatening to shake off theory in one way or another, threatening to explode understanding with an affective gesture, as it slides around from one possibility to another, doing what it does best, doing what it can.

What remains to be considered is what holds an image apart from the performing body - i.e., the *actor* - that turns up to support the image, to do the actions, to put over the words in the way they need to be put over. No two actors slide, or pass out of sight, in the same way. In the modicum of difference, or we might say better iconoclastic refusal, that each actor brings to the performance of their act, in the minimum failure that accompanies any appearance, the image of any of us that is forged out of all of our labour in the state of exception is held up to view and put under pressure. *Something else is done*, and maybe too *something else* appears - some intensity of appearance that we may find ourselves laughing about later, even if we don't *know* what it is we are laughing about or what it is we *do* when we laugh like that.

REFERENCES

Agamben, G. (2004) *The Open: Man and Animal*, trans. Kevin Attell, Stanford: Stanford University Press.

Agamben, G. (2005) *State of Exception*, trans. Kevin Attell, Chicago: University of Chicago Press.

Brecht, B. (1984) *Brecht on Theatre: The Development of an Aesthetic*, ed. and trans. J. Willett, London: Methuen.

Burt, R. (2008) 'Preferring to Laugh', *Parallax* 14.1: pp. 15-23.

Felman, S. (2003) *The Scandal of the Speaking Body: Don Juan with J. L. Austin, or Seduction in Two Languages*, trans. Catherine Porter, Stanford: Stanford University Press.

Jameson, F. (2000) *Brecht and Method*, London: Verso.

Read, A. (2007) *Theatre, Intimacy and Engagement: The Last Human Venue*, Houndmills and New York: Palgrave Macmillan.

Ridout, N. (2007) *Stage Fright, Animals and Other Theatrical Problems*, Cambridge: Cambridge University Press.

Body Events and Implicated Gazes

JIM DROBNICK

To raise the issues of appearance in performance art immediately engenders its antithesis - disappearance. As Peggy Phelan articulates in one of the canonical texts of performance studies, beneath the vernacular understanding of the term 'performance' lies a philosophical and psychoanalytic quagmire. The ontology of performance, she concisely formulates, is representation without reproduction. That is, it is the nature of performance to make visible representations, or constructions of the real, but in a way that resists reproduction or commodification by the 'smooth machinery' of capital. The 'maniacally charged present', which performance awakens into being, also posits a sense of the 'now' that is deeply disturbing to conservative cultural values (Phelan 1993: 146-8). One such conservative value that I will address in this essay bears a particularly interesting and problematic relation to the staging and experience of performance, especially bodies on display. It regards the appropriate manner of viewing 'art' - that is, the aesthetic gaze.

Amidst the proliferation of discourses surrounding the body and performance in recent years, the problematics of appearance have been central to discussions of the body's meaning, representation, agency and identity. Since the theorization of the gaze by Sartre and Lacan, numerous critiques have undermined what at one time was held to be the noblest and most trustworthy of the senses. The unitary, objective status of vision has splintered into innumerable fragments according to various ideological biases and exploitative practices: the male gaze, the colonial gaze, the tourist gaze, the medical gaze, the penal gaze and so on. It should come as no surprise, then, to consider the aesthetic gaze as an extension of this list.

The detached, disembodied, disinterested form of artistic contemplation is a form of looking that is hardly natural but one suffused with ideological values. As just one example of such values emerging in contemporary criticism, David Boland dismisses artists' use of their own bodies because it is an unmediated encounter that can place the viewer in only one position - that of a voyeur. Without aesthetic distance, he asserts, it is impossible to make judgments about the work. We become moved, not by choice or volition, but through 'visceral fascination', much like the response to pornography (Boland 1995-6: 42).[1] The threat Boland experiences at being confronted with a living presence is palpably evident in his writing. Not only are Christian concepts of shame and sin implicitly raised in his comments but so too are the legal and pathological determinations of vice and perversion. That artists may intentionally address the very conditions and problematics of such viewing practices is tellingly absent from his analysis.

Pierre Bourdieu calls the structure of perception that Boland nostalgically advocates the 'pure gaze', a stylized manner of concentration upon artworks as if they were complete unto themselves. The aesthetic attitude embodied in the pure gaze is not a self-evident,

[1] This uneasiness with the implicated gaze afflicts the critical reception of performance art in general. See Taylor (1998) and Ward (2001) for further examples of critics seemingly despairing at the loss of a stable, transcendent and authoritative viewing position.

Performance Research 13(4), pp.64-74 © Taylor & Francis Ltd 2008
DOI: 10.1080/13528160902875648

spontaneous occurrence but a product cultivated through a particular apprenticeship between art consumers and art institutions:

> The experience of the work of art as being immediately endowed with meaning and value is a result of the accord between the two mutually founded aspects of the same historical institution: the cultured habitus and the artistic field. Given that the work of art exists as such... only if it is apprehended by spectators possessing the disposition and the aesthetic competence which are tacitly required, one could then say that it is the aesthete's eye which constitutes the work of art as a work of art. (Bourdieu 1987: 203)

The implication of Bourdieu's argument is that definitions of art, artist and aesthete evolve in a concerted, mutually-influencing manner. The aesthetic or pure gaze, if it is to persist, can only be reproduced as long as there is a tacit agreement, or implicit collaboration, between artist and art viewer.

It is this process of interrogating this tacit agreement that, in my view, accounts for the particular power of performance art, which is predisposed towards framing and heightening the issues intrinsic to the activity of viewing. Whereas the pure gaze may objectify and aestheticize artworks which are (for the most part) designed to be objectified and aestheticized, difficulties begin when that gaze is directed towards breathing individuals, even when that gaze is openly invited. The performances I will discuss here, while involving the aesthetic gaze, force a re-evaluation of distanced, disinterested contemplation. Instead, they cultivate a gaze that is embodied and interested, in other words, an *implicated* gaze.

I use the word 'implicated' in contradistinction to 'embodied' although these terms are, in many ways, closely related. Even as my line of thinking is indebted to critiques of vision, such as that by Jonathan Crary who in *Techniques of the Observer* noted a shift from a pure to physiological understanding of opticality in the nineteenth century, my use of the term 'implicated' is intended to address the spectator in the process of looking a few degrees past the state of embodiment. 'Implicated' connotes notions of complicity, involvement not only physically but ethically. Boland, for instance, acknowledges the embodiment of vision in his critique of performance art but fails to recognize (and indeed reacts unconsciously to) his own implication in the process of display. Rather than feeling absorbed into a spectacle, these performances compel an extra-conscious realization that one is engaged in a precise and concrete activity of viewing.

BODY EVENTS

Artworks that highlight the implicated gaze fall under a category of performance that I designate *body events*. Body events are a conceptual model for re-evaluating the manifold relationships between the practices of looking and the politics of the body. A thumbnail definition might read as such: the body placed in an heightened viewing matrix so that both elements - body and view - are integrally connected. In body events, the experience of the spectator is itself a primary element for contemplation and critique. The mode of aesthetic apprehension is not assumed to be *a priori* or unproblematical, but is simultaneously solicited and challenged, called forth and denied. Body events treat the aesthetic gaze as an aspect and material of the artistic process, a practice to be engaged, reworked and reflected back upon itself. In short, body events disturb assumptions of a natural and transparent viewing position and seek to uncover its political and ideological investments - they corrupt the pure gaze by activating the implicated gaze.

Artists engaging with body events inherently appropriate and dialogue with display practices in and out of the art world. Erasing the distinctions between the domains of art and self, between artist and art object, where the aesthetic gaze's ideology of beauty and commodification is made evident and countered, has characterized the works of Vanessa Beecroft, Gilbert and

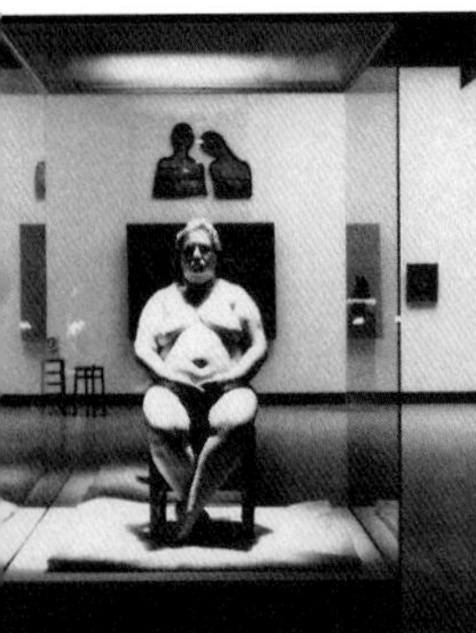

• **Peter Greenaway, *The Physical Self*, 1991, views of installation and performance;** *photos: courtesy the artist.*

George, Hannah Wilke, Sylvie Fleury and Chris Burden. For Luigi Ontani, Cheri Gaulke, Scott Burton, James Coleman and Pierre Joseph, the format of the tableau vivant is resurrected as a means to address the stereotyping gaze and how it factors into mythic constructions of nationality, gender, sexuality, history or cultural icons. Body events may also focus on practices of discipline, self-mortification and ritual, and the works of Linda Montano, Marina Abramović, Tehching Hsieh, Stelarc and Joseph Beuys implicate the viewer in both mystical and highly physical manners. The colonializing gaze, one that manufactures questionable and denigrating forms of difference, is confronted through works by Coco Fusco, Guillermo Gómez-Peña, James Luna and Rebecca Belmore. Activist works, such as those by Mona Hatoum, Suzanne Lacy, the Women of Greenham Common, Joey Skaggs and ACT UP, place vulnerable bodies on display in sensitive or symbolic public spaces so that they cannot be ignored by a business-as-usual populace. And the voyeuristic gaze is both seduced and challenged by Bob Flanagan, Annie Sprinkle, Felix Gonzalez-Torres and Carolee Schneemann, where issues concerning pleasure, erotics and the sex industry are brought into the open and debated.

In a culture where the objectification of others is normal and accepted, even profitable, the act of self-objectification, which many of the above works actively engage, is a powerful technique for interrogating and transgressing convention. Self-objectification creates a dissonance in the harmony of cultural consensus because, in the words of Phelan, it 'suspends the proprietary relation between body and being' (1995: 204). By exacerbating this gap between body and being, body events demonstrate the contingent and malleable nature of identity. In the sections that follow, I will examine five body events created by artists who utilize or evoke the implicated gaze not only to critique ideologies that try to constrain and overdetermine the body's relation to being but also to explore alternative and affirmative possibilities.

PRIMALITY

Several of Peter Greenaway's curatorial projects have incorporated the display of live persons, sometimes in association with the filmmaker's cinematic production. *The Physical Self* (1991), the first of such displays, raised the possibility of engaging with a pure, aesthetic gaze, and then, just as quickly, demonstrated its impossibility. The exhibition stationed unclothed models in glass vitrines, which were interspersed among sculptures, photographs and installations that foregrounded aspects of the body culled from the collection of the Museum Boymans-van Beuningen in Rotterdam. Set up as archetypal exemplars - such as 'Man', 'Woman', 'Age', 'Mother and Child' - these living displays intended to manifest (and celebrate) the pristine, fleshly materiality that exists prior to the vicissitudes of artistic training, formal stylistics or critical interpretation. For Greenaway, the individuals provided a phenomenological 'origin' for art:

> These figures are the markers, the templates, the basic models, to which all the paintings, sculptures, drawings and artefacts of the exhibition relate. To them we can compare, evaluate and correlate all the imaginative adventures that the collection houses. (1991: 11)

Through the use of the unclothed figures as 'reference points', the exhibition thus situated the act of viewing between the opposite poles of the creative act: between the artist's gaze upon raw material and the connoisseur's gaze examining the finished product.

There is a subtle anxiety caused by the individuals on display here not only because the contrast of real bodies (silent and immobile) against representations of bodies in artworks (charged and expressive) may appear to privilege the latter at the expense of the former. The live bodies serve to anchor what is indisputably a diverse selection of artworks, from all historical periods, as well as to provide an element of facticity to alleviate the abundance of conflicting aesthetic styles and intentions. It is through the attempt to arrive at a primal, corporeal substrate that precedes the relativity of figural representations that its opposite is, in fact, demonstrated: materials are never completely raw (uses are foreshadowed in the very process of manufacturing) and artworks are never finished (but are renewed, reinterpreted by novel contexts).

Despite removing the clothes and the particularities of identity, class, biography, community affiliation and the like from these bodies on display, signifiers such as hairstyle, pose and musculature link them inextricably to the social. The insulating enclosures of glass, however ordinary and ubiquitous, nevertheless espouse an ideology of transparency, where knowledge is open to all and nothing can be concealed. In the effort to bring us as close as possible to the minimal presentation of the self, Greenaway problematizes that which is normally taken for granted - the purity of vision. Even at the zero-degree of performance, at a point hypothetically prior to representation or the artistic act, the framing of the gaze is paramount. The viewer, held between contrasting domains of the artist and the connoisseur, the real and the representational, the transparent and the impenetrable, is forced to recognize how each gaze is implicated in the other.

The ideal of primality is also queried by the inevitable presence of the curator's own voice and intentionality, or, in other words, *parabasis*. Martin Jay, in his examination of the possibility of writing an 'original', self-sufficient and impersonal text, i.e., one without citation, name-dropping or pleas for legitimation via the use of external authorities (in many ways a project similar to Greenaway's), argues that such an original text can never be achieved because of the operations of *parabasis* - the intrusion of the authorial voice. Even in a seemingly objective text, parabasis is evident, for the voice is impossible to suppress completely (Jay 1996: 25-6).[2] The bodies on display in *The Physical Self*, however expressionless, silent, stripped and motionless, ultimately testify to the agency of the curator: that each model was chosen, hired, positioned, scheduled, remunerated. Just their very presence speaks of, as well, institutional decision-making, notably budgetary negotiations, security concerns, public relations rationales, health protocols, contingency plans. As much as these bodies are poised to exemplify a primal 'origin', something archetypal, unmediated and beyond history, they inevitably speak of the power and authority exercised at a particular place and time, with a certain group of individuals, for a specific curatorial purpose.

RENEWAL

The aesthetic gaze is physically performed and the object of display in IRWIN's contribution to the Arts Festival of Atlanta, *Transnacionala* (1996). IRWIN, a collective of artists hailing from Slovenia, mock the heroic posturings of the early avant-garde. Suspended by guy wires about 3 metres above the floor, the artists beheld an exhibit of their own paintings - parodies of religious icons and avant-garde paintings - attached to the ceiling. The simplicity of the ninety-degree shift in viewing practice nevertheless unleashes a provocative series of allusions. Looking upward, of course, is the trope of transcendence and heavenly identification prominent in Christian vault mosaics and the *trompe l'oeil* ceiling decorations of the Baroque and Rococo periods. This up-ended viewing perspective wittily mocks the utopian claims of modernist vanguardism that a 'revolution in perception' - as promised by successive waves of stylistic 'isms' - would affirmatively transform societal values.

It is this last point that matters most for IRWIN who, in the words of one of the collective's members, 'revives the trauma of avant-garde movements by identifying with it in the stage of their assimilation in the systems of totalitarian states' (Cufwer and IRWIN 1994: n.p,). Despite the appearance of defying gravity, in reality their postures render them hopelessly

[2] Jay mentions a second trope, *prosopopoeia*, which will be discussed below in relation to Danny Tisdale's work.

• **IRWIN, *Transnacionala*, 1996, performance stills;** *photos: courtesy Mary Jane Jacob and the artists.*

controlled and vulnerable. The romanticization heaped upon the avant-garde in the West is mixed with cynicism and suspicion in Eastern Europe. *Transnacionala* concretizes the term 'scopic regime' and demonstrates how viewing can be an imposed practice, a dictatorial bodily discipline. At the same time it mocks avant-garde iconography and demeanor, for the 'new' weightless, horizontal mode of viewing art is clearly arduous, absurd and irrelevant to aesthetic experience.

Transnacionala also involved a cross-country excursion to meet and confer with U.S. artists, hold screenings and conferences and generally attempt to communicate across cultural, social and political differences between the first and second worlds. This bears remarkable similarity to that form of tourist/educational practice of the eighteenth and nineteenth centuries, the grand tour, only reversed. Whereas wealthy English and North American youths sought to acquire the sensibility and aura of the old world, IRWIN seeks (or at least performs a seeking of) similar knowledge upon new world territory.

The ideal of renewal, so fundamental to the early-twentieth-century avant-garde as well as to a culture at large enamored by technological advances and the potential of a new century, not only inspired artistic practice and creation but also served as its subject matter. The turning point, or *peripeteia*, that IRWIN references is a constant theme in avant-garde writing and manifesto production since its origin in the 1820s. While diatribes against tradition and stylistic inertia are common to almost any era, it was in the first decades of the past century that the call for renewal fulminated into an attack on all aspects of lived experience. This radical turn sought to institute ways of life unbound by habit, culture and even physical laws. All aspects of society were to be transformed: the environment rebuilt, behaviour restructured, the body redisciplined or transcended, the senses reinvigourated. By rotating the posture of viewing, IRWIN recalls those efforts by industrial engineers, educational reformers and government officials, as well as by artists, choreographers, composers and playwrights, to remold the body according to new societal demands (for efficiency, health, national spirit) and to express a new vocabulary of artistic meaning. Hillel Schwartz identifies 'torque' as the kinaesthetic paradigm upon which much of this rethinking of the body was centred (1994). IRWIN's torquing of the spectating position concisely sums up the utopian longings infusing these bodily transformations, as well as their utter impracticality.

Thus far, my discussion of IRWIN has evolved from a definition of *peripeteia* as a kind of turning, but the term also implies a reversal of fortune, one intimately tied to *anagnoresis*, recognition, a shift from ignorance to knowledge (Aristotle 1961: 72-3). For IRWIN, a form of this reversal and recognition occurred at that time when the utopian aspirations of the avant-garde and their desire to revolutionize the nature of everyday life were appropriated by the political formations of fascism, Nazism and Stalinism. IRWIN's painfully awkward stasis, hovering between heaven and earth and belonging to neither, is telling. 'Liberated' from gravity but subject to physical duress, 'free' to enact a new posture but hopelessly bound by its enabling apparatus, the members of IRWIN symbolically stage 'the avant-garde' at its moment of *peripeteia*, but before *anagnoresis* has transpired. Hanging, in limbo, seemingly caught in a trap of their own devising, the performers physically demonstrate that precarious point between the idealism of the avant-garde's intentions and the recognition that these ideals can be (and have been) translated into totalitarian practices.

IRWIN's twisted viewing perspective no doubt also confirms the enormous difference between the artistic and social histories of the east and west. Can work be transported from one context to the other and expect immediate comprehension? Eastern European art, when noticed by western observers, has traditionally been denigrated as derivative of western artistic movements and symptomatic of the degree to which non-democratic cultures repress individuality and free expression. That this Cold War rhetoric persists in contemporary critical discourse can only be the result of an unwillingness to engage with a different cultural tradition and aesthetic dynamics. What is the proper way for westerners to view Eastern European art, synecdochically represented by IRWIN's ceiling-mounted paintings? IRWIN cagily avoids a direct answer, but their performance indicates that it is a gaze based on subtly dissimilar postures and principles.

ALTERITY

'There's a shortage of the "other",' quips the Dean of Anthropology at the Museum of Ex-Minorities and Ex-Savages. 'It's simply supply and demand' (Tisdale 1994: 24) As fictitious as this curator is in Danny Tisdale's book *Danny, The Last African American in the 22nd Century*, he nevertheless adopts that particular confluence of discourses - oriented towards education and entertainment but immersed in the exploitative politics of racial difference and the salvage paradigm - so characteristic of the ethnographic gaze. The book is a companion to the artist's performance in his installation *The Black Museum* (1992), a collection of 'artifacts', i.e., commodities marketed to African American consumers, that both perpetuate and challenge stereotypes. Danny, displayed on a pedestal, wears an 8-ball jacket and is flanked by his own image on the cover of popular magazines. His subjection to the ethnographic gaze references not only the sensationalizing blend of fact and fantasy that typified P. T. Barnum's and William Bullock's exhibitions of non-Western peoples but also individuals such as Ishi and Ota Benga who really had no other choice but to submit themselves to denigrating public display. While the narrative explanation for Danny's appearance two hundred years in the future (a dentist's mistake in sedation) may strike a fabulist note, the reason for him being the last of his kind (ethnic cleansing wars) evokes a violence that is still an all-too-real occurrence.

Danny's solution to the problem of objectification is to pattern his career after athletic and free-enterprise models. As the 'First Anthropological Free Agent' and president of Kulture, Inc., a 'human display company', his motto is 'I exploit, before being exploited' (Tisdale 1994: 24 and interview with the author 1996). He thus simultaneously inhabits the dual and opposing roles of curator and artifact curated, entrepreneur and promoted commodity, spokesperson and celebrity specimen. The audience's gaze may retain elements of exploitation, yet it also becomes a means of Danny's personal enrichment and empowerment. While his story articulates a response to and a critique of his unusual predicament, an option not readily available to his historical counterparts, it also implicates our gaze in such a way that we cannot ignore how the museum functions as a legitimizing context of display. Danny's persona exemplifies a critique of the museum as much as racial prejudice and the exoticized representations of non-white peoples. Whether or not Tisdale's 'Museum of Human Display' will come into being anytime soon, or has similarities to any specific existing museum (it recalls many nineteenth-century examples), it serves as a useful fiction underscoring the fact that transcendental notions of truth and knowledge have less influence in authorizing the museum's pre-eminence than its attention to market forces and sensationalist desires.

The consumptive nature of the ethnographic gaze relies partly upon the silence of those displayed, of their not being able to counter projections of inferiority, primitiveness or other ideologically motivated constructions of difference. Locating individuals in the realm of myth rather than in history evacuates their credibility as a voice in current debates and their ability to exercise agency in contemporary politics. As Danny implies, control needs to be taken not only of one's economics but also of one's meanings. If display contexts in the past

• Danny Tisdale, *The Black Museum*, 1992, performance still; *photo: courtesy the artist.*

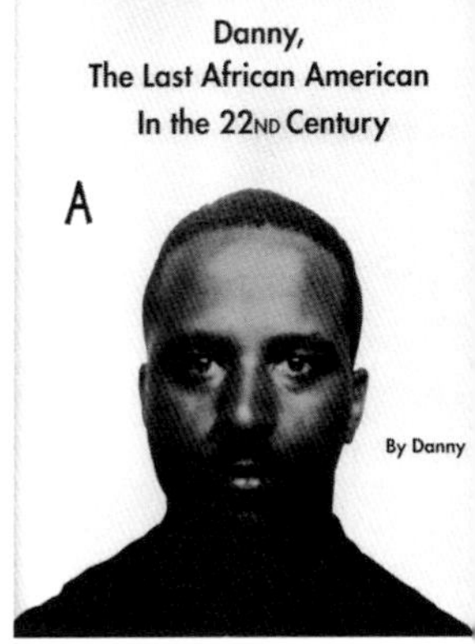

• Danny Tisdale, cover from *Danny, The Last African American in the 22nd Century*, Atlanta: Nexus Press, 1994; *photo: courtesy the artist.*

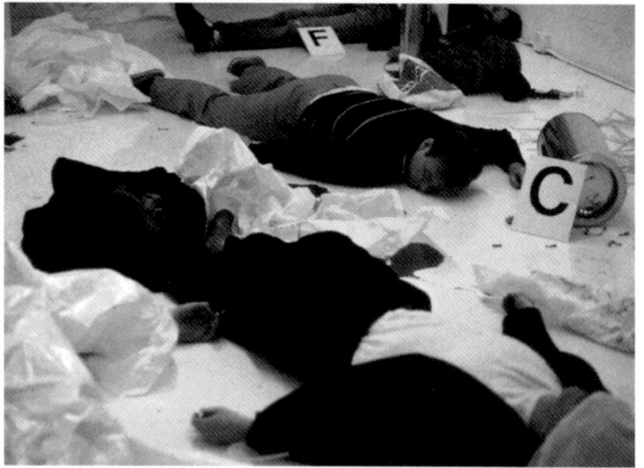

• **Graham Durward, *Rome Airport* 1986, 1992, performance stills;** *photos: courtesy the artist.*

reduced persons of non-white status or non-Western origins to mute objecthood as artifacts or specimens, Danny is a living example of *prosopopoeia*, or personification. He is the artifact that speaks, that tells its history in no uncertain terms, that indicts its audience in the very moment of informing them of his story. Danny is the artifact that resists projection, conjecture and assimilation into racist or pejorative discourses by the fact that his speaking dispels the very notions of a long lost origin, anonymous production or mysterious use (in other words, the obscurity of the artifact) that supports and necessitates the explanatory intervention of experts. While Danny's degree of control may still be subject to the power and influence of reactionary discourses, his testimony provides at least one example of an incontrovertible, countervailing presence. As the oratorical skills and writings of ex-slaves in the nineteenth century disproved racist assumptions of Africans being ineducable and uncivilized, Tisdale again utilizes voice, text and visibility as the means to defy stereotypes and achieve self-definition.

REPETITION

If the horror of death in Danny is implied, Graham Durward's *Rome Airport 1986* (1992) directly engages its spectacular nature. Referring to the bombing mentioned in its title and the photograph of the appalling aftermath published worldwide, showing corpses, strewn baggage and police markers, the artist hired actors to assume the dress and positions of the victims. Audience members walked through the scene of carnage and could inspect the painstakingly duplicated props scattered across the floor of a New York City gallery. Loud airplane landing noises defeated attempts at conversation, forcing viewers to be silent and introspective. While the effects of the violence is here made palpably evident, the tenor of the performance is disturbingly ambiguous. Is it outraged at or sympathetic to the act of terrorism? Critical or complicit in how the media portrays and reproduces images of brutality? An ironic swipe at artists who rhetorically valorize the criminal persona and aesthetic 'transgression'? The link between art and death, beauty and murder, championed by de Sade, De Quincey, Bataille and others, where violence is romanticized as a 'pure', inescapably 'real' act, recurs in this performance. While terrorist bombings in Europe were extremely visible events in the American media in the 1980s, Durward performatively staged what had been rarely experienced on American territory but had generated reams of commentary. Predating the World Trade Center and Oklahoma City attacks, the context of this performance was to some degree American 'innocence'; now that the experience of terrorism is horrifyingly first-hand, *Rome Airport 1986* assumes the status of a rehearsal, a trial run to test the emotions and response to such a calamity.

Regardless of the moral positions one could take on the meaning of terrorist acts, the viewing position invoked in *Rome Airport 1986* collapses several distinct forms of looking pertaining to both images and bodies. One might argue that the performance confirms the commemorative attitude prominent in recent 'anniversary culture'. However, the performance is less a simulation or a reenactment, such as one would find in a living history museum, than an attempt to re-corporealize an image. The bodies on display here participate in a moebius-strip-like loop of mediation, in which a historical event is processed through a series of representations with shifting ontological levels, subjective attachments and domains of reference. From bombing to photograph to performance and again to photographic documentation, the event that gave rise to *Rome Airport 1986* is less of a presence than the structures, and gazes, that attend to it: the mourning relative, the shocked bystander, the scrutinizing detective, the news-gathering reporter, the condemning US State Department representative, the querying art patron. The mute, stilled bodies instantiate the panoply of gazes and viewing positions that focus on this site to produce what, in other contexts, will be called criminal violence, political revolt, sensational news, inhumane evil and even (need it be said?) art. The physical violence documented in the photograph foreshadows a more subtle violence to follow - via interpretation, opinion, spin-control and propaganda - in which the event is commandeered to substantiate any number of positions.

Durward's piece critiques the fantasy of presence not through entities such as the simulacra, Baudrillard's copy without an original, or the hyperreal, Eco's term for that which is more real than real, but a self-conscious repetition, or *palilogy*, that makes no pretense of seamlessness or believability. Viewers are not necessarily intended to marvel at the performance's likeness to the bombing photograph, to be duped by its illusion, to be fearful of its uncanniness or to be shocked anew at the horror of what had by then become a depleted media image. Rather than attempting to appropriate or supersede the original (not only the photograph but the event itself), the work positions itself as yet another in a series of significations, like others yet also distinct, a repetition that is different by not presuming to be anything other than a repetition. For the artist, acts of terrorism are 'meaningless', that is, they gain significance from the proliferation of discourses attending to their aftermath (Durward 1997: interview with the author). The strength of the performance lies not in its ability to mobilize judgement of the morality or political effectiveness of such acts, but to be immersed in the process of meaning-making. Like Taussig's notion of 'mimetic excess' (1993: 254-5), which is distinguishable from mimesis proper by its self-reflexive character, Durward's restaging/repetition attempts to revisit that moment in which significance is produced and to render that moment available for reflection. If the psychological allure of statues lies in their receptivity to projection, the temporary statues/corpses in *Rome Airport 1986* are poised to call forth and secure for witnessing that very process of projection. While an air of relativism may hover over this work, reinforcing the truism that the meaning of events (and artworks) is contrived rather than inherent, it nevertheless underscores anew the understanding that the construction of meaning is itself a performative process.

CONSUMPTION

If there is a seductive or glamorous sheen to the photograph attracting Durward's attention, it is a glamour quite inadvertently produced. Trudie Reiss, on the other hand, working at a similar intersection of photography, mass media, the body and performance, investigates glamour that is intentionally and industrially generated. Her 'directorial' performances consist of staged photo-shoots, where fashion models are arranged and documented - the difference being that rather than given instructions to smile, be natural, radiate charm or use any of the other tactics conducive to creating the allure of 'beauty', they are intimidated into crying. Even though the models are professionals hired specifically for each event, the emotional relationship is cruel and difficult to watch. As Reiss abusively intones 'Imagine coming home and finding your children have all been killed ... Look at you, you're not getting any younger ... You don't have a career, look at what you're doing for a living', the models display the entire spectrum of sadness: from introspective doubt and listless melancholy to sniffling and violent bawling. Whether or not the crying is acted, manipulated or genuine, the women in *Two Pretty Girls Crying with Camera* (1993) and the male model in *The*

Crying Man (1994) clearly are positioned as victims of the systemic objectification and violence endemic to the manufacture of ideal images.

The intensity of Reiss's appropriation of the role of the fashion photographer lies not so much in deconstructing the romanticized mythos of the photographer, of exposing the brutal 'reality' behind the facade of glamour, even though it does both. Its importance lies in performing the psychodynamics of the fashion gaze, that is, in combining *pathopoeia* (arousing emotions) with *phanopoeia* (making visible). It is not necessary, for instance, that her enactment refers to any specific (male) photographer because it refers to a more pervasive societal gaze that insatiably fabricates and elicits consumption of 'loaded' emotions to sell, command attention and influence opinion-making (Reiss 1997: interview with the author). Being a witness to the abusive treatment of models disrupts manners of facile visual consumption routinely associated with magazines and other vehicles of the seduction/ spectacle industry. In other words, Reiss's performance arouses emotions, makes visible systemic abuse and basically taunts us to intervene. If the aura and authority of art make that intervention a slim, even absurd, likelihood, an ethical challenge is nevertheless raised. As Heidi Gilpin theorizes about the symbiotic relation between psychoanalysis and performance, the recreation and re-experiencing of a traumatic event is a means by which an individual can take an active role in relation to that event; it is a method for survival and healing (1996: 110). Witnessing the suffering that Reiss's models seem to experience thus enacts a cathartic reversal; it is as if the pain they are berated into exacts a symbolic retribution for the millions of consumers who identify with the glamour gaze but have internalized its misogyny, who admire supermodels yet hate themselves for never measuring up. By staging the photographic gaze, Reiss crystallizes glamour into an artifact that is constituted equally by revulsion and hate as it is by adoration and seduction. In this theatre of (emotional) cruelty, the identities of victim and victimizer are shifting and troublesome.

• Trudie Reiss, *Two Pretty Girls Crying with Camera*, 1993, performance stills; *photos: courtesy the artist.*

For Stuart Culver, the manikin (i.e., model) stimulates consumption by 'project[ing] the image of a complete body while simultaneously dramatizing a present lack' (Culver 1988: 107). By mediating the relations between consumers and commodities, manikins compel people to confuse themselves with objects (110). Reiss's performance lies precisely at this point of mediation and highlights that consumption is hardly a neutral activity, one without psychic costs. At the same time, the artist submits this mediation to a deft reversal, from one of identification and pleasure to alienation and pain. Reiss's performance involves ambivalence - as the abusive language humiliates, the camera glamorizes. The photo-shoot thus utilizes a two-pronged approach - discursive and optical - to create a contradictory field of praise and humiliation.

Far from being an aberration from the excessive devotion fuelled by the showbiz industry, Fred and Judy Vermorel argue that hostility towards celebrities and cultural icons is a 'necessary consequence of unconsummated, unconsummable passion' and is endemic to the systemic provocation provided by the 'frustration machine' of pop (1985: 249). Reiss's bullying and sadism, while clearly outside of the narrow bounds in which the entertainment and advertisement industries wish to channel consumers' behavior, nevertheless instantiates widespread and under-recognized attitudes towards the figures cultivated to promote desire. Viewers in Reiss's scenario thus stand in a strategically uncomfortable and awkward position. Caught in a conflict between glamour and mistreatment, they cannot just simply

consume the proceedings, and yet they cannot quite intervene - it is a tense stalemate between appearance and emotionally devastating affect.

ACTS OF RECONFIGURING

Unlike critiques that seek to undermine the hegemony of vision through a metaphorical violence to the eye, by splintering or decentring the gaze or through some kind of fusion, creating synaesthetic perceptual faculties such as 'tactile vision' or technologically enhanced 'mobilized' gazes, body events critique the look, especially the aesthetic gaze, through a doubling process that intensifies and compounds it. The distance and detachment that serves as the foundation for the asymmetrical power relations of the aesthetic gaze is undermined by the face-to-face encounter between performer and audience. This 'doubled gaze', a looking compounded by being looked at, sets up an intersubjective relationship, a reciprocal engagement, that Jennifer Fisher argues 'confound[s] the privileged viewing normally operative in display culture' (1994: 31). The gazes between viewers and performers not only intersect with each other but also overlap with other cultural examples of viewing. The body events of performance art inevitably (and intentionally) allude to spectating practices normally considered outside of the realm of art, such as voyeurism, surveillance and exoticization, yet in such a way to force the structure of vision to work against itself in a self-reflexive, destabilizing manner.

Any human display inevitably embodies questions about the politics of relationships, in other words, it provokes an ethical conundrum. As Allan Kaprow notes, looking, or focusing attention, is a form of taking a stand, an exercise in 'moral intelligence' (1993: 88). If one agrees with the tenet that every display engages in some form of objectification, viewers are placed in a no-win bind in regard to their activity: to look is to be complicit with the objectification, not to look is to deprive oneself of participation in the world. The performances discussed above force viewers to acknowledge their own complicity in objectifying viewing practices, yet they also provide a framework in which to negotiate their problematics.

Critical anxiety over 'visceral fascination' contains an element of validity beyond its intention to dismiss body art from serious artistic consideration. Body events' blatant use of the visceral and the implicated gaze are not an end in themselves, but are strategically employed and redirected towards examining the politics of viewing. The uneasiness experienced at witnessing and being drawn into a viewing relationship (that to some may seem perverse, immoral or illegal) is a sure sign that socially-approved viewing patterns are being unsettled. That one may feel out of control and beyond volition is an accurate perception: the assumed rules of visual behaviour have been deliberately contravened.

In these performances it is not possible to naively or confidently adopt the pure, aesthetic gaze because the ethical dynamics inherent to that type of viewing are exposed and raw. As Catherine Elwes notes, once the distance and safety of the voyeur's 'cloak of invisibility' have been removed, viewers are exposed to the consequences of their own desires (Forte 1990: 263). The implicated gaze refuses to allow viewers the option of hiding their responsibility for the act (or performance) of viewing behind habitual cultural rationalizations. The exoticizing gaze, the consumptive gaze, the sensationalizing gaze - these are objectifying gazes legitimized by various cultural principles: the 'pleasure' offered by the entertainment industry, the 'free speech' espoused by advertising campaigns, the 'information' transmitted by television news media. The aesthetic gaze, while ingrained with its own objectifying principles, is permeable enough to allow artists to combine and contrast one gaze against another. Body events reveal these gazes without (or contrary to) their structuring, legitimizing framework, where we are forced to experience them in all of their ethical complexity.

The nexus of terms in which Phelan locates the ontology of performance - representation and reproduction - accounts for the particular ephemeral effects of performance. The implicated gaze that characterizes the body events discussed above, however, exceeds the binary terms of her distinction, partly because it concerns the nature of performance not so much as it vanishes or resists commodification but as it heightens and affects experience in the brief moments of its occurrence. Performance, as Phelan theorizes, unquestionably represents (or not) and reproduces (or not), yet performance also possesses a dimension that reconfigures. By reconfigure I mean that performance creates a charged relationship in which conventionalized behaviours and roles are not only made visible but are also made available for re-imagining and refashioning. Rather than conforming to the paradigmatic bifurcation of the viewer and the viewed, with all of its attendant ideological baggage, the body events outlined here incorporate viewers in an uncertain and unpredictable social engagement, one grounded upon the ethical processes inherent to appearance. The implicated gaze triangulates the terms of representation and reproduction by acknowledging the viewer's and artist's embodiment, along with the experientiality of the space's mise-en-scène. If the implicated gaze questions the foundations of social propriety and forces a confrontation with an audience's positionality, it is only so that the cultural politics of the body can be exposed and potentially reconfigured.

REFERENCES

Aristotle (1961) *Poetics*, trans. S. H. Butcher, New York: Hill and Wang.

Boland, David (1995-6) 'Body Art: Cheap Thrills,' *Art Monthly* 192: 42.

Bourdieu, Pierre (1987) 'The Historical Genesis of a Pure Aesthetic', *Journal of Aesthetics and Art Criticism* 46: 201-210.

Cufer, Eda and IRWIN (1994) 'NSK State in Time' in *IRWIN: Geography of Time*, Umag, Croatia: Dante Gallery Marino Cettina.

Culver, Stuart (1988) 'What Manikins Want', *Representations* (Winter): 97-116.

Fisher, Jennifer (1994) 'Exhibiting Bodies: Articulating Human Displays', *Border/Lines* 31: 18-31.

Forte, Jeanie (1990) 'Women's Performance Art', in Sue-Ellen Case (ed.) *Performing Feminisms: Feminist Critical Theory and Theatre*, Baltimore, London: Johns Hopkins University Press, pp. 251-69.

Gilpin, Heidi (1996) 'Lifelessness in movement, or how do the dead move? Tracing displacement and disappearance for movement performance', in Susan Leigh Foster (ed.) *Corporealities: Dancing Knowledge, Culture and Power*, New York and London: Routledge, pp. 106-128.

Greenaway, Peter (1991) *The Physical Self*, Rotterdam: Museum Boymans-van Beuningen.

Jay, Martin (1996) 'Name-Dropping or Dropping Names?: Modes of Legitimation in the Humanities', in Martin Kreiswirth and Mark Cheetham (eds) *Theory Between the Disciplines: Authority/Vision/Politics*, Ann Arbor: University of Michigan Press, pp. 19-34.

Kaprow, Allan (1993) *Essays on the Blurring of Art and Life*, ed. Jeff Kelley, Berkeley: University of California Press.

Phelan, Peggy (1993) *Unmarked: The Politics of Performance* (New York, London: Routledge.

Phelan, Peggy (1995) 'Thirteen Ways of Looking at Choreographing Writing', in Susan Leigh Foster (ed.) *Choreographing History*, Bloomington and Indianapolis: Indiana University Press, pp. 200-10.

Schwartz, Hillel (1992) 'Torque: The New Kinaesthetic of the Twentieth Century', in Jonathan Crary and Sanford Kwinter (eds) *Incorporations*, New York: Zone Books, pp. 70-126.

Taussig, Michael (1993) *Mimesis and Alterity: A Particular History of the Senses*, New York, London: Routledge.

Taylor, Diana (1998) 'A Savage Performance: Guillermo Gómez-Peña and Coco Fusco's "Couple in the Cage"', *TDR* 158 (Summer): 160-80.

Tisdale, Danny (1994) Danny, *The Last African American in the 22nd Century*, Atlanta: Nexus Press.

Vermorel, Fred and Vermorel, Judy (1985) *Starlust: The Secret Fantasies of Fans*, London: Comet Books.

Ward, Frazer (2001) 'Gray Zone: Watching *Shoot*', October 95 (Winter): 115-30.

Appearance, Reality and Truth in Magic
A personal memoir

aladin

I have an uncertain relationship with the term 'magician', its typology and its connotations in the popular imagination. My own practice as magician and/or interdisciplinary artist (the terms for me are interchangeable) is however much influenced by early exposure to itinerant conjurers across India, Pakistan and Bangladesh.

In what we know as the Asian subcontinent, nomadic and tinker (here in the sense of 'make-shift', 'prone to improvising') subcultures or castes of entertainers are likely to have originated as early as Indus, Dravidian and related civilisations of approximately 3-4,000 BC. In childhood I recall being cognizant of the perpetual journeying that marked magicians apart from those who plied other trades - and it was unspoken that *there have always been magicians*. Incidentally, in my early days magicians were always approved for encounters in yards, gardens, on the pavement - but never allowed in the home.

Although I am not all that clear about being a magician, I have had my share of magical epiphanies.

In my earliest years, perhaps when I was close to two, my father from time to time quite spontaneously (and in hindsight quite opportunistically) devised feats of magic which to me seemed suited to, if not arising from, immediate circumstances. In fact my earliest memory of magic per se dates back to when we were living in Geneva, watching with growing astonishment as my father, unhurriedly and

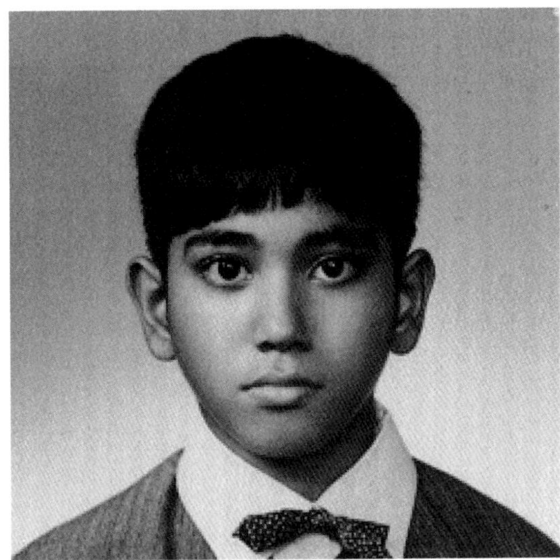

without preparation, *invisibly* 'transported' some quite unwieldy toys of mine into a light fixture suspended from the ceiling at a distance from us. I heard each toy land on the other as he slowly and systematically 'flung' every one into 'storage'.

Then one night, in the most impossible of conditions and at the risk of his life, my father extemporised the complete disappearance of our family and all our personal effects from our home in Baghdad - only for us to reappear in London. This was at a crucial stage of the Bangladesh independence war and the extraordinary illusion allowed my father to go on to play his part in that country's history. Fittingly, years later, this epic of misdirection featured in a National Geographic television documentary.

Performance Research 13(4), pp.75-81 © Taylor & Francis Ltd 2008
DOI: 10.1080/13528160902875655

When we were living in Paris I was once struck by how my father and a fellow Sufi looked into rather than at each other - it seemed *unconscious*.

Another time, in Karachi, I saw a street-boy of about eleven dig up a coin from every pit in the ground that had been created by my repeatedly flinging a stone as far as I was able to.

In Geneva I saw a clown besieged and befuddled by a disc of light which in turn pursued him and was pursued by him.

We were an itinerant family. I learned to entertain myself, for we moved so frequently from country to country; in the course of which I became self-taught and self-sufficient in the arts of prestidigitation. I wasn't yet ten when I gave my first performance on a proscenium - in Kolkata (Calcutta); having trained in classical Indian music, on some of those occasions I would appear as a singer alongside billing as a magician.

My cultural heritage encompasses the three Asian countries mentioned above; my mother was from Dhubri, Assam in the East of India and my father was primarily from Kishoreganj, East Bengal (which became East Pakistan and then Bangladesh) but also had Syrian ancestry. The traditional - classical, folkloric and popular - live arts were engaged with matter-of-factly; most families regardless of class, caste or circumstances were at a minimum informal makers of culture - as amateur artists, wordsmiths, musicians or actors etc. with their families and friends as audiences at the very least.

My early experience of entertainment and live arts was thus of seeing them sited in the community and being generated and sustained by real relationships. Of course, this profoundly coloured my awareness and practice of the arts. In adult life too my schooling has continued: I have had almost 20 years' parallel experience of strategy consultancy alongside carrying out street-work with vulnerable people. The latter went hand-in-hand with continuous professional training in psychotherapy, counselling, group work and brief intervention skills applied to itinerant, public contexts. I recognise how this sustained engagement with youth and community development has an influence over the choices I make as an artist, both in terms of subject matter/content but more particularly in terms of choice of means.

I have my own theories about magic.

Conjuring or magicianship appears to invoke an agency to instrumentalise matter and processes such that their characteristics conflict with scientific norms.

In practice, the instrumentalisation is effected through literal if disguised means, entailing devices or dexterousness - or some combination thereof. I hardly need add that some quarters posit that actual agency is also involved, as opposed to mere dissembling.

As a practitioner of magic, I am myself reliant in the first instance on purely prestidigitatory means to anchor 'illusion' - alongside modes of communication such as speech, gesture, observation, call-and-response, and so on. Like many before me I have developed my own plastic, sleight-of-hand vocabulary. I find these tools so much less cumbersome (and certainly more malleable) than a 'box of tricks' - leave alone a company of assistants and a trailer truck of equipment. It's therefore not much of a negotiation for me to realise a transformation to say, 'perambulating legerdemain'; I don't even need the seclusion of a telephone box!

There is an obsession with 'appearance' and 'disappearance' in popular discussions of magic. For what it's worth, however, conjurers' own

Photo: Andrew Atkinson

lexicons would recognise perhaps a dozen categories of illusions demonstrating the ability to turn the laws of nature on their heads. Here is just one of these typologies, helpfully synthesised by a friend who is both a fine artist and a magician: 'Production, Vanish, Transposition, Restoration, Penetration, Levitation, Animation, Suspension, Mind-Reading (or Clairvoyance) and Physical Anomaly (i.e. Headless Woman)' (Sheridan 2002: 25).

I thus see my practices as a magician (and strategy consultant) to be concerned at their core with illumination and revelation. At the same time I do feel resistance to being drawn into a discussion about the techniques including technologies of 'deception' or 'fabrication'. All this focus on *means* is disproportionate and seems to imply that magic can be understood through knowledge of its *techniques*.

Perhaps at its apogee magic transcends its means. For my part, my intention in conjuring is to incite as well as to *instrumentalise* (for a purpose other than gratification) a transcendental state in the 'audience', as opposed to devising an intervention whose sole purpose is to *elicit kudos* for any imagined or real virtuosity on the part of the magician. Of course the very *presence* of a magician on its own has been known to trigger this very state of quasi-psychosomatic enchantment.

I have said that I instrumentalise the magical experience. What I mean is that in my unfolding and developing practice my formal interventions are curated and devised to engage the 'audience' in a wider narrative. I am interested in the discourses preoccupying civil society - and as an interdisciplinary artist my actions are inevitably emerging from and responding to events around;

Photo: NicholeRees

my hope though is that my actions as a magician or otherwise are dialogic as opposed to didactic.

Nothing like being preached to by a magician!

Mainstream, public notions of magic sometimes pose challenges for me in my own practice.

My creativity as a magician is, no doubt, to some degree spurred by my frustration with public, and indeed creative community, characterisations of the live art form of 'magic' to essentially be about the exposition of virtuosity at creating illusion (or its semblance) - of extreme dexterity and/or marvellous ingenuity, sometimes embellished or underpinned by a nominal narrative structure. If I have not already made that point - I also find limiting and frankly bathetic the labelling of magic as exclusively a tributary of popular entertainment.

The art form does have at its functional, working heart some of the processes of what we can crudely call 'deception'. However, the latter's more pejorative connotations as well as the high/er profile of popular/populist conjuring have contributed to a qualification, if not the *tainting*, of magic's claim to be considered a 'serious' live art form, whose practitioners are capable of having discursive, critical theoretical preoccupations. It has its complexities if one has things to say as a magician but are not seen to fit into the continuum between con artist, deity/public icon, children's entertainer and so on.

One also needs to take account of the impact of the near-global peer networks of magician-entertainers (most of whom are passionate amateurs).

The above manifests itself in the visibility, numbers and public profiles of 'successful' magicians who achieve their standing through evidence of popular following or knowledge, or sometimes through fame engendered by the mass media; below the tip of this iceberg are a host of individuals who are almost without exception self-selecting members of a plethora of organisations for magicians, aspiring magicians and their associates, who are in turn in close engagement with practitioners deemed to be 'successful': all this creating its own critical mass to iteratively derive and legitimise typologies of magic and magicianship.

These loose agglomerations of magicians define and patrol the mainstream discourse on magic - and erect canons and epistemologies which through their (at the very least tacit) endorsement by formal and informal peer groups give rise to hegemonical definitions of magic and magicianship which are not necessarily as comprehensive or inclusive as I have been indicating. Indeed these definitions function as an ideology which is not amenable to easy interrogation due to the sheer asymmetry of power and demography, whereby mainstream practitioners are preponderant and thus weigh heavily in the debate such as it is.

It all makes it difficult to define and *declare* oneself a magician if one feels one's practice does not conform to these mainstream formulae and if one is aware of not sharing the same knowledge base.

To put some perspective on the challenge of contesting magical orthodoxy, one can draw a hypothetical analogy with music. Were it left to popular musicians (by definition the most populous sub-set of practitioners of music) to promulgate a definition of 'music', there would be scant chance of ethnographical or anthropological considerations receiving prominence; however no-one would argue that these latter were illegitimate, irrelevant or even *marginal* on the basis of being a strictly minority view.

However, as stated earlier, the practices and following of *popular* magic by-and-large eschew critical, theoretical or dialectical enquiry and do not generally value it as a legitimate area of concern. Hence the popular and popularly maintained online encyclopaedia Wikipedia offers a taxonomy of magic and magicians which has long privileged 'entertainment' as a reference point and classed magicians as performers of 'tricks' and 'effects'. It is a predictably reductionist and exclusive view of magic - which extends to excluding from classification or inclusion those magicians who may not have 'peer' standing or validation, as I sketched earlier above. Their own entry about me is a helpful case in point - my relative invisibility on the conventional magicians' radars and Richter scales a past source of ambivalence and tendentiousness on the part of authors/editors.

As I hope I am making clear, I identify myself as a magician whose inclination is for more metaphorical landscapes than the (more literal) ones of engendering deception and illusion for their own sakes. Whilst historically it has been the *virtuosity* alone of the conjurer that has been the primary object of gaze, in my own excursions as a magician (and live artist, as one and the same) I have been attempting to place virtuosity and critical enquiry side-by-side, each to leaven and interrogate the other.

In my continuing attempts to elucidate the proscriptions and conventions of *popular* conjuring and its *prevailing methodology* - wherein the event or action is prescribed to be a form of divertissement, more often to draw attention to the semblance of or actual virtuosity - I am at the same time clear in my mind that an *authentic, encompassing* contemporary magical eschatology/A-Z would and should admit a practice of magic which can be *effacing of virtuosity, and which subordinates it to critical purpose.*

I find it valuable to devise expositions of 'magic' to inhabit either/equally spaces associated with contemporary interdisciplinary arts practice or public spaces that are in every sense unmediated.

Although my preoccupations are discursive and critical, in my 'actions' I do draw extensively from a palette in which virtuosity features prominently (however subordinated). It makes for a dichotomous relationship - one that stretches between the populist divertissement of 'dazzling sleight of hand' and the strictly experimental, improvised, essentially un-premeditated and avant garde; at times - even often - I do feel *conflicted.*

Ideas for multifarious projects arise from my abiding interest in the processes of civil society - with which I also have a professional engagement with a hat on as policy adviser and strategist consultant. Germane also is my recognition early on that the persona of 'magician' was more or less afforded licence by society at large to enter otherwise proscribed territories.

As my experience of magic has evolved, so I have become more alive to its catalytic (always instrumentalised) properties - for instance when 'dropping a magician into unexpected contexts'.

In the ten years to 2008, I embarked on a tranche of interventions in the public domain, away from formally designated 'cultural venues', whereby I more or less outwardly effaced my inclinations to subvert whilst at the same time (Trojan Horse-like) seeking out 'engagements' which would permit experiment/s, covertly or otherwise. Several of these resulted from direct approaches to me - through the website I maintain as a 'Magician' - from individuals or organisations wishing to contract an entertainer. It is in an *inquisitive* (though occasionally inquisitorial) spirit that I foisted these live/public art actions-by-stealth on (mainly unsuspecting) patrons.

Closer to a mainstream centre-ground, during 2007-8 I directly interrogated what it is to be a magician in a time of unsustainable global consumption and production practices - undertaking public artistic collaborations with individuals, organisations and also U.K. government agencies including: a Texan banker, a community health group, the British Museum, the Department for Environment, Food and Rural Affairs (DEFRA) and the Department for Culture, Media and Sport (DCMS).

The collaborations immediately above succeeded an accidental 'magician in residence' project which took place 2000-4. It was a period of being appointed to public office in London government in my capacity as a magician/artist - to contribute to the development of a culture plan for the capital city; I was Co/Vice-Chair of the Cultural Strategy Group at City Hall.

In 2008 I embarked on a number of conversations with practitioners across sectors with the aim of assembling a suite of actions to take place in the unmediated public space. Provisionally titled 'Throw', the intention was for the project to unfold over some years as a series of one-to-one collaborations between myself and about a dozen others.

At the core of 'Throw' would be the *virtuosity*, physical and emotional rituals, processes and actions involved with throwing a playing card at some considerable speed (over 100mph) and distance (the better part of a hundred metres). Prospective collaborators would take responsibility for bringing and binding intentionality and purpose to the spectacle; each obliged to devise, prescribe, curate, direct and above all site the actions which I would then have to carry out. My primary task in all this would be to identify and secure collaborators. The 'outcomes' are not conceived as formal theatrical events requiring production and communications support.

I have long toyed with collaborating with a neuropsychologist to map and describe 'the morphology of the magical event', particularly its incontrovertible effect on our physical and emotional processes.

There are three images, created specifically for this piece, that I collaborated in.

They are by the photographers Andrew Atkinson, Nichole Rees and Marcus Tomlinson. They were taken in the photographers' own studios, two days apart in the same week in April 2008. *None of the images has been digitally altered.* One of the photographers utilised natural light only, another was at pains to light me with studio lights and the third experimented with multiple techniques in situ to subvert my appearance whilst shooting *straight*. Unexpectedly I became struck by how these images illustrate the paradox of how we each see what we want/need to see - neatly encapsulating my digressions above about how the magic I know and see is not necessarily that which is discerned/discernible to others.

REFERENCES

Jeff Sheridan, Jeff (2002), in 'Conjuring and its Cousin', in Helen Varola (ed.) *Con Art*, Sheffield: Site Gallery, pp.25.

Main photo: Marcus Tomlinson

Theatre and the Technologies of Appearances
The spirit of apprehensions

ANTHONY KUBIAK

> Without things,
> There would be no appearance or disappearance;
> Without which,
> There would be no things.
>
> Nagarjuna

1.

At the age of eight, while walking through Peckham Rye, William Blake beheld 'a tree filled with angels, bright angelic wings bespangling every bough like stars'. He saw 'haymakers at work, and among them angelic figures walking' (Bentley 2001:20); he spoke to the dead, and when still a young child, saw the face of God peering through his bedroom window, which set him shrieking with fear and confusion. Another time he 'saw the Prophet Ezekiel under a Tree in the Fields'. He spoke to the dead King Alfred, held concourse with prophets and angels, and moved easily in the celestial realms. Blake's visions were Real events - intense, detailed, and above all *coherent* - and lasted throughout this life, although he apparently realized that others did not share his propensities. In *Vision of the Last Judgement*, Blake famously declared:

> I assert for My Self that I do not behold the outward Creation & that to me it is a hindrance & not Action; it is as the dirt upon my feet, No part of Me. "What," it will be Questioned, "When the Sun rises, do you not see a round disk of fire somewhat like a [coin]?" O no, no, I see an Innumerable company of the Heavenly host crying, "Holy, holy, holy is the Lord God Almighty." I question not my Corporeal or Vegetative Eye any more than I would Question a Window concerning a Sight. I look thro' it & not with it. (92)

Blake's deeply held belief in the Imagination - for him the realm of truth and the Real - was held in esteem by the later Romantics, although they clearly thought Blake himself insane ('There was no doubt that this poor man was mad,' wrote Wordsworth famously). Although holding to Blake's notion that Imagination was the seat of creative and transformative power, they were never quite ready to cross the line into the spirit world with him. Even his colleague, John Varley, an astrologer, was only willing to move within a spiritualism that merely glimpsed the 'other side' from the safety and comfort of the quotidian world. For the Romantics, Imagination was still, above all, a psychic realm, a kind of interior dreamscape, impinging upon, but separate from, the exterior world of what we now call 'consensual reality'. For all of their repudiation of neo-Classical thought, the Romantics were already, after Blake, deeply impaired by neo-Classicism's empirical vision of the world. And little has changed since then.

Indeed, it is astonishing that still today, after 200 years of Blake criticism, his visionary[1] consciousness is still taken as pro-visionary, less than actual: that for all of the extravagance of his experience as a Moravian shaman, Blake's world is still far more than we imagine, can imagine, or, apparently, wish to imagine. And while his contemporaries thought him mad, we, who have ensconced Blake's later emissaries in the pantheon of visionary modern artists - Artaud, Lautremont, Rimbaud, and Yeats, Rothko, Messiaen, Beuys and Kazuo Ono - are still not

[1] It is important to note here that by vision or visionary I do not necessarily mean visual. The word 'visionary', although it contains the visual root, so to speak, also suggests the theoretical, the theatrical and the performative, as well as the speculative, and creative agency in general.

Performance Research 13(4), pp.82-92 © Taylor & Francis Ltd 2008
DOI: 10.1080/13528160902875663

quite sure what to do with *his* visions, the very *ecstasis* that made Blake Blake, and made his contemporaries, no slouches among the epoch's *decadents*, blanch. Today, within the contexts of neuroscience and empirical explanation, we might suggest reasons or causes for Blake's visions: temporal lobe epilepsy, hypnagogic episodes, other brain or neurophysiologic disease - which, though possibly true, do not explain his hold on us, our attraction to the visionary impulses that he awakens in us, the desire to move in worlds other than our own, not through mere desire for escapist fantasy, but in search of some truth - the truth of what, in the theatre, we understand as psychedelic (literally, 'mind manifesting') reality.

I invoke Blake's memory, then, for a double-purpose: to suggest that a particular and critical mode of performative consciousness operates both within what we think of as canonical literatures and art, as well as on the fringes of culture. This critical performative mode is, moreover, precisely unlocatable, and before we ascribe this 'unlocability' to some poststructural mode of thought, I would insist that this critical performative mode, this visionary impulse of mind embodies the best hope of curative, redemptive *political* life. I will, moreover, call this mode of consciousness variously visionary, shamanic, or *pharmakeic*. Now, I use the somewhat creaky Platonic/Derridean signifier here precisely because of its overdetermined unsignifiability. *Pharmakeus*, a critical term in Plato's *Phaedrus*, means many things and all of the following: wizard, sorcerer, magician, physician, poisoner, and, finally, scapegoat (more literally, the *pharmakos*). The term is 'apropos' in the present context because it implies the visionary and shamanic, the curative, as well as the source of dis-ease. It also implies, as Derrida points out in his famous essay 'Plato's Pharmacy', an exclusion, a kind of ghostedness through which the missing term leaves its trace in the text. Derrida says that Plato never uses the term *pharmakos* in the *pharmakeia-pharmakon-pharmakeus* chain of signifiers in the text of *Phaedrus*, but that it exists palpably as an exclusion. So the term itself becomes a *pharmakos* - an exclusion or expulsion. As such, *pharmakeus*, like the ghostly *pharmakos*, operates as a kind of parasitic vector, both a poison and a cure, a kathartic agency that moves what is within the community's heart outside the city walls, or within the body outside the body - vomit, sweat, shit - a kathartic agency whose imago is the caduceus, the doubly-twining/twinning serpents, the promise of cure raised up in the midst of desolation and sickness, a promise of both life and death.

Finally, the *pharmakeia-pharmakon-pharmakeus* trope situates the problematic term 'shaman' as well: much of the discussion of shamanism over the past forty years has centered on the term itself - whether it is or should be specific to Tungus culture, to which the word owes its origin; whether such a thing as 'shamanism' exists as a cross-cultural phenomenon; whether the shaman is anything more than a romanticized idea combining a fascination with indigenous culture and magic with extravagant claims to esoteric knowledge and practice. Countering these objections are others that see shamanic agency as a kind of marginalized spiritual healing practice and a type of political and cultural resistance, not unlike the function of the artist in our own culture - combining the roles of analyst, seer, and prophet. The word 'shaman', like the figure of the shaman itself, then, is fraught. It represents a concept seemingly known cross-culturally, but not identifiable in any final or complete way - it marks a locale but not a one specific role, it is, in a word the *phamakos* again, the curative agency of exclusion and expulsion itself. And here I would also acknowledge the fact of poststructural disappearances as well (see my apologia for Derridean terminology above): what is now oftentimes regarded as a retired dead-end theoretical trope is itself excluded in the historiographic and postcolonial analyses of current theoretical bias - but in fact poststructuralism represented, among other things, the last possible insinuation of other modes of thought into Western, materialist analysis: poststructuralism, in the work of Deleuze, Lacan, Derrida and others (none of whom thought of themselves as poststructural, with the possible exception of Derrida), but also tantric, Vedantic, shamanic, and kabalistic thought. The exclusion of poststructural thought, is, in other words, the theoretical *pharmakos* of our recent past, and represents the exclusion of an authentic Other, the Real remnant of indigenous, non-dualistic, non-materialities.

It is by now clear, I think, that I do not intend to discuss Blake at length here (Blake himself now a becoming-*pharmakos*), but, rather, simply wish to position his visionary oeuvre as a kind of conceptual boundary, and to suggest, through it, a theory of thresholds or boundaries that might help us gain access to world views and modes of cultural consciousness currently antithetical to the nearly universal presumption of material practices and the appearances they supposedly entail. I am suggesting this theory of thresholds as *pharmakon* to a current critical praxis (Deleuze and Lacan on the one hand, and Badiou and Laclau on the other) that hovers along the performative fault lines of thought, the fractures or cracks that will either outrun the course of materialist ontologies, or confront the dematerialized necessities of post-Capital ennui. This is precisely the sort of cultural crisis, anthropologist Michael Taussig reminds us, within which shamanic or *pharmakeic* thought thrives and transforms its sociopolitical world. Moreover, Taussig suggests, this transformation is wrought, in part, through a kind of dematerializing Artaudian theatre of cruelty:

> the predominance of the left hand and of anarchy – as in Artaud's notion of theatre of cruelty with its poetic language of the senses, language that breaks open the conventions of language and the signifying functions of signs through its chaotic mingling of danger and humor. (Taussig 1987: 442).

The theatrical grounding of these transformative modes are extended, in Tausssig's thought, to Brecht's theatre of the strange as well, bringing to the idea of shamanic performance a decidedly political edge. Taken as a whole, Taussig's view of shamanic practice aligns with the concerns of political theatre, the performative impulses of cultural resistance realized through the very cultural traditions that gave it (cultural resistance) birth.

This view runs counter to the views of performance theorists of recent age – Schechner, Turner, and others – who, though they have pointed out the social efficacy of shamanic practice, often presume that such worlds are effectively lost to us but in certain flavours of marginal performance,[2] or performances which are *merely* socially effective, thus gutting the central, spiritual impulses of shamanic practice. This limited and limiting perspective has overshadowed the import of visionary politics as a fertile response to the impoverishments of materialist/capitalist theory. This invocation to a visionary politics is, in fact, Taussig's thesis in *Shamanism, Colonialism, and the Wild Man*, which suggests that, through overcoming the terrors visited upon them in the rubber baronies, and then re-embodying those same terrors through shamanic practice, the *curanderos* of Amazonia, though losing the innocence of pre-colonial life, restored their own power and authority by partially overcoming and absorbing the depredations of colonial exploitation through the Imaginary worlds of a visionary and spiritual shamanism. Within the very terms of the shamanic practices of the *curanderos* he visited, Taussig shows how they (the shamans) took the poison of colonializing terror within themselves, absorbed it, and then expelled the same terror back into the colonized minds of the white men who were exploiting them. This is a mirror of the shamanic ceremony of the *vegetalistas*, in which medicine is taken in, medicine which is also a poison, medicine which induces dis-ease, and then expelled in order to open up a visionary world that will provide psychic and political salvation. Blake himself drew this doubled *pharmakeus* in the life serpent – bringer of life

[2] It seems that whenever the shamanic/magical approach to art is invoked, one is obliged to name works which either exemplify it or negate it. I do not wish to do that here: whereas Schechner may see an 'authentic' shamanism in certain modes of performance (Anna Deavere Smith [!]), but deny its presence within the classical canon, for example, I choose to think of the shamanic as culturally specific to the degree that it is often precisely *in* the canon (understood in its widest sense) one can find the impulses to shamanic, *pharmakeic*, mind. By the same token, inasmuch as shamanism is generally thought of as a more marginal type of spirituality than religion, one can also find its impulses in the most marginal and underground types of art and performance. The *Pharmakeus* is performance specific.

and damnation - coiled about the reclining body of mother-Eve, and also in the bodies of life-giving, destroying angels moving up and down the heliacal Jacobian ladder of dream in a prefiguration of Michel Serres' own ambivalent angels of information, moving along the coils of DNA (Deoxyribonucleic acid), within the virtual parallel channels of computerized imagination and sluiced through the Imaginary worlds of gamers and theorists, and into the realms of the techno-shamans of the informational age. Here the specific content of Blake's political resistance is less important than the deployment of visionary mind to these ends - in Blake, it is the visionary process itself that is crucial, rather than the content of the visions themselves. Psychologist Benny Shanon, writing of his experiences in the Amazon drinking the *pharmakeic* medicine, ayahuasca or *yajé*, about which I will shortly say more, writes this in his *The Antipodes of the Mind*:

> Ayahuasca visions reflect neither what is hidden in the mind of the drinker nor mind-independent, Platonic-states of affairs, but rather, they are the works of creation. While the recourse to creation [as a possible explanation of the substance of ayahuasca visions] salvages us from Platonic realism, psychologically it is much more radical than may seem at first glance. (Shanon 2002: 396).

We are in essence, Shanon suggests, creators. Our minds are as-if minds, metaphorical minds, artistic minds, and beyond the forms of creation, the agency of that creation provides us with new worlds, unthought possibilities, and - quite separate from the content of those worlds - in the very substance of mind, the possibility of possibility itself. This is, in itself, not a new idea: Karl Jung, Theodore Roszak, and James Hillman among others in this century were the impetus to an engagement with magical and shamanic states of consciousness. Though the examples they invoked were often classical and Romantic ideals, and not modern ones, the inspiration behind the move of performance theorists of the sixties and seventies into the shamanic mind is often traced back to these and even earlier thinkers like Liebniz, Schopenhauer and even Kant.

What is unusual here is that recent critical theory - in the likes of the late Foucault, Derrida, Deleuze and others - has found its own, presumably materialist, limits at something like the nexus of philosophy, spirituality and aesthetics, and seems to invoke, knowingly or not, the magical and the visionary beyond the limn of signification - this 'beyondness' again reflects the Derridean sense of *pharmakos*, the doubly excluded middle. This impulse to the magical is, then, not simply born of paralysis or exhaustion - there is no simple retreat into phantasy or opiation here. Rather these limits recapitulate the desire of an earlier Marxist sentiment, in writers like Georg Lukács, to remake the world, a world in which, to quote Michael Tucker paraphrasing Lukács, '"transcendental homelessness" could only be overcome and the "integrated totality" of life and meaning achieved once more through the eventual world-wide historical triumph of the working class.' (1992: 29) Exchange 'indigenous wisdom' for 'working class' and you have something of the desire today to go back to shamanic consciousness as the means of revisioning the world. Indeed, when we consider earlier approaches to shamanism in performance, we notice the tendency among the emergent anthro-theorists to posit shamanic-ritual as origin, as that which precedes the later, rationalized and conceptualized 'theatre'. We see in these early valorizations of (shamanic) ritual over theatre the presumption of shamanic practice as preparatory to the theatre, as original (closer to origins), and, protestations to the contrary, primitive. But if we de-posit theatre as the result of a conventionalized ritual performance, and see theatre as precisely the space within which ritual and magical appearances are thought, then the distinctions between shamanism and theatre begin to blur. Shamanism might then be seen as a type of theatre preceding ritual, and both theatre and shamanic practice can be understood as elaborations of a more primal (though not primitive) visionary impulse to reformulate the world.

Thus while Lukács and Marxism see the means to this end in the reformulation of the means of production or 'material practice', the desire underlying the artistic impulse is the same. And in a world in which the meaning of materiality is itself up for grabs, the pursuit of visionary consciousness and a concomitant theory of thresholds seems less like nostalgia or escape than true revolutionary thought and desire. Put more simply, in the prosaic words of Jimmy Weiskopf, a journeyman in the yajé visionary world, 'Until you drink *yajé* you simply do not understand that all of us are literally full of shit and for that reason cannot see beyond the material realm.' (Weiskopf 2005: 40). We certainly see this anti-materialist revolutionary impulse in the work of Joseph Beuys, in the trance encounters of Karen Finley, and the jazz flights of Bill Frisell, or the cruel, transcendent butoh of Kazuo Ono. And beyond this, in the work of Taussig and particular modes of critical thought, we see the taking in of the poison-medicine, its transmutation through and into thought, and its expulsion as writing. There is a letter from the interior waiting on the nightstand when we arrive home, preparing us for dream and vision, calling us to the apothecaries of mind and the healing prophecies of un/consciousness.

2.

Throughout the sixties and much of the seventies, a psychotropic Dreamtime held sway in American society both through the onslaught of global pharmo-capitalism and narco-terrorism, and the psychonautic extravagances of LSD-25, psilocybin and mescaline, and later, in the intensities of Ecstasy, crystal-meth, and heroine (chic). Eventually the drug culture was seen - outside the purview of 'legitimate' pharmo-industrial treatment - as self-numbing, self-medicating mass-opiation and narcissistic self-absorption. Seemingly countering the progressive politics of the American left in the sixties and seventies, the drug culture absorbed and obliterated the warning cries of revolutionary epigones now gone to ground. It was even rumored, among the political hard-core, that the Keyseyian Acid-tests themselves were but an outgrowth of the Central Intelligence Agency's MKUltra experiments with hallucinogenic drugs, which were designed, finally, to render insensate the counterculture and its Antiwar Movement after earlier efforts to produce psychedelic uberwarriors failed the test. In any and all cases, the single universally reviled aspect of the sixties culture was - among those on the political left and the right - the use of hallucinogenic drugs toward spiritual and visionary ends. Whatever shamanic impulses remained were taken over by the pale visage of libertarian politics and New Age drumming practices, aromatherapies, and crystal-gazing which promised (and failed to deliver) the ecstasies of spirit without danger to the body/mind. At the same time, as apropos the framing of consciousness in the Modern age, the emergent cultural materiality of the academic left drove the Symbolic roughshod over the *élan vital* of the Imaginary, Imagination and Imaginal modes of critique.

Some pockets of resistance remained, however. Within the strange post-Vietnam silicon culture of Southern California,[3] the last remnants of the hallucinotactical rear guard of sixties psychonauts were incubating electric dreams of global consciousness through their Eniacs and Commodore 64's, giving birth to the computerized episteme of the Information Age. Here, within the interstices of midnight programming sessions and late night peyote trips into the desert (to which Michel Foucault readily assented) the alchemical quests continued, and, through the earlier research of Richard Schultes, the great Harvard botanist, and later on, in the travels of psychic pioneers like Terrence and Dennis McKenna, within the retorts and crucibles in Alexander Shulgin's dusty lairs or the practices of mystical pioneers

3 For a compelling account of this period, and the elision of post-acid culture with the rise of Silicon-Valley techno capitalism, see Eric Davis' *TechGnosis: Myth, Magic & Mysticism in the Age of Information* (New York: Three Rivers Press, 1998).

of the tantric mind, the visionary impulses of sixties alchemical culture was kept alive, ultimately directing scores of burned-out cultural materialists to find their souls in the Great Apothecary of the Amazon basin.[4] The various quests, though politically and ethically questionable according to some,[5] have borne interesting fruit, not the least the discovery of the medicine of the Amazonian *vegetalistas* and *curanderos*, the literal elixir of life, ayahuasca.[6]

This particular liana, ayahuasca (banisteriopsis caapi), appears ubiquitously in the Amazonian shaman's marketplace. When pulverized using wooden mallets and boiled, along with other native plants - chacruna, toé, and chaliponga (psychotria viridis, various species of brughmansia, and diplopterys cabrerana, respectively) - this mixture eventually forms a thick and bitter-sour tasting medicine called ayahuasca, or *yajé*. Ayahuasca, the medicine, is, then, also the local name of the vine, banisteriopsis, that is its principle ingredient.

The 'vine of souls', or 'vine of the dead'[7] grows in a double-helix, like other tropical vines, two liana in grotesquerie wrapping about each other in a vegetal-serpentine slow-dance. Some popular authors[8] have described its shape as visually reminiscent of the DNA helix, and sometimes as the caduceus, twin serpents entwining a healing staff set to ward off disease in the wilderness. Without overtaxing the intriguing chiasmus of DNA models, ayahuasca, and caduceus which is, after all, purely accidental, I present this fortuitous convergence to suggest a larger, more global and emergent synthesis that is still unfolding today - a convolution of image, symbol and mutated consciousness that appears in the work of Blake and continues on in the later visionaries of modernism, through the realms of visionary and imaginal art in general, art that suggests the shaman-doctor's power and acumen, and functions not unlike ayahuasca as a cultural emetic, inducing katharsis in its partakers as a means of cleansing both body and soul.

The vine itself is of pharmacologic interest because it is an MAO inhibitor, which means that it contains compounds (harmine, harmalin) that hinder certain enzymes from breaking down substances before they can be absorbed into the bloodstream. The MAOI properties of banisteriopsis work with psychotria - a plant whose active hallucinogenic ingredient N, N-DMT would normally be destroyed by stomach proteins - and the very potent brughmansia or datura, to produce an extraordinarily intense hallucinogenic sacrament that is at the heart of indigenous shamanic and spiritual practice. Interestingly, it is quite common for those taking ayahuasca - even non-indigenous people who have never seen the twining growth of banisteriopsis - to have visions of coiling snakes - the reticulated anaconda, specifically - as well as visions of the jaguars and the huge and spectacular butterflies that haunt the Amazon rainforests.

The effects of the medicine are profound: after initiating its powerful purgative effects (vomiting, diarrhea, profuse sweating), the visions begin. Underwater worlds of snakes and river gods, celestial realms of alien and angelic beings, silent music and ethereal dancing, dead ancestors and the Jungle Mother herself may appear to give inspiration, direction and instruction. Christ may appear, or Moses, even if one is a staunch non-believer, or perhaps the demon if such a thing is necessary to the advancement of shamanic, *pharmakeic* wisdom. The visions are often frightening and can be extremely unpleasant, evolving slowly into the terrible and the wonderful. Terrible and wonderful sometimes become indistinguishable in vision, and the medicinal world of *yajé* sometimes seems more real than the world one leaves and to which one returns. Other worlds appear and are entered, uncanny beings encountered, songs heard, messages conveyed with a level of tactile reality unlike even the most powerful of the hallucinogens of previous decades. Unlike LSD-25 or mescaline, ayahuasca produces very real visions. In fact partakers often resist the term 'hallucination', believing the worlds visited are the true realms, while the

4 All overshadowed by the unfortunate and ego-driven media-man Timothy Leary.

5 The somewhat predictable objections of cultural exploitation have arisen as more and more Westerners go to the Amazon for visionary healing. The fact seems to be, however, that the influx of money, far from ruining the indigenous culture, has revitalized it, and has brought increasing attention and respect to the shamanic traditions of the Amazon Basin.

6 The current rise in the domestic popularity of the medicine is somewhat curious–it is in no way recreational, nor is it uniformly or even generally pleasant. The use of the medicine will, I suspect, self-select the users who come to it.

7 Both translations of the Quechua word *ayahuasca* are accurate.

8 Narby, Jeremy (1998) *The Cosmic Serpent: DNA and the Origins of Knowledge*,New York: Tarcher/Penguin.

quotidian world is debased and illusory. (Comparisons to Artaud's theatre of cruelty are, again, hard to avoid, as are the world views of tantric practice.) The effect of ayahuasca is powerful and profound. Few who take it within its sacred contexts walk away unchanged.

The 'discovery' of ayahuasca by Western anthropologists and psychonauts in the middle part of this past century brought several curiosities to light apart from the peculiarities of the medicine's effects themselves: how was it, for example, that out of the numberless plant species in the Amazon basin (many of which have not yet been catalogued or even recognized), the local *vegetalistas* or shamans were able to puzzle out the synergistic action of these two plant substances, and actualize them together (one cannot simply eat banisteriopsis and psychotria together to experience their effects). And why would anyone, after the typical intitial terror of ayahuasca vision, take it a second or third time? Or more importantly, how it was that this medicine - for that is what ayahuasca is - has so effectively molded and shaped Amazonian cultures to its own ends? What are we to think, for example, of the documented action of the medicine that produces, as I've already suggested, similar imagery among different drinkers, and consistent iconography and similar visions across multiple cultures? When one considers the commonalities among the visions, one wonders - apart from the mechanism (either brain peptides or phytochemical interactions) that might account for them - what do the shared images, narrative and visions suggest about cultural specificity? If a medicine opens one to becoming-anaconda, for example, is that culturally specific or not? Considering the centrality of snake iconography in Amazonian culture, are my dreams my own or the effect of acculturation? What if I have snake-dreams while drinking the medicine here, in Irvine, instead of the ceremonial hut in Iquitos? What if I drink with no knowledge of the origins and cultural contexts of the medicine? Here the cultural boundaries break down, as they do in the *yajé* dream itself. The possible cause of these oddly syncretic, synergistic and serendipitous improbabilities remains more than mere mystery. The cause becomes the question of causality itself - the origin of the world in mind, the relationship of consciousness to environment, history and aesthetics. The causes and questions are the same causes and questions, as Taussig suggests, animating theatrical appearances themselves, reflecting life, reflecting mind. And they are in no way framed solely within the Artaudian theatres of cruelty and excess, as Taussig himself notes in his yoking of Brecht to the visions of *yajé* nights. Brecht, Taussig suggests, like Lukács, was interested in remaking the world, and felt it could only be done by overturning the structures of capitalist impulse and desire that fragment the world:

> Epic theatre aimed not at overcoming but at alienating alienation, twisting the relationship between the extraordinary and the ordinary such that the latter burns with a problematizing intensity in a world that can no longer be seen as seamless and whole. (1987: 329)

What both Brecht and Lukács miss, however, as do many post-Marxist thinkers, is that the problems emerging from an alienated and shattered world are not due to capitalism per se, but, rather, to what gave capitalism its birth, the presumptions of materialism that lock us into a single mode of consciousness and which manifests as the empirical, historical, and material. Taussig suggests a different model, one based upon the shamanic modality of consciousness in which everything is also its opposite. Taussig tells a story in which Manuel, the fifteen-year old son of a white man, takes *yajé* and has a powerful vision of the devil, appearing in the place of the shaman, 'But it wasn't the devil, it was the shaman. It was he who had been the devil,' he says. (325) The shaman who had been the devil who had been the shaman appears in an infinite regress of appearances and their opposites, echoing, once again, the ways in which the shaman takes upon himself the projections of the white man, and reflects them back in a mirroring of the terrors that white culture has thrust upon the indigenous cultures of Amazonia. Weiskopf, on the other hand, has a somewhat more idiosyncratic view of the transformative effects of the medicine; 'It is only,' he says,

> when you have been forced to your knees, completely helpless, and feel the vomit flowing from one end of your body and the shit from the other that you will have scaled the first rung of the ladder to heaven. (2005: 45)

Heaven, we presume, being liberation from the material delusions that, as he says earlier, fill us with shit in the first place, but a liberation, as Artaud reminds us, also demarcated by humility and acceptance; the world will one day collapse around us, to paraphrase Artaud, and the theatre, the visionary, is there to remind us, first of all, of that.

3.

If there were a single theorist alive in the past fifty years with whom Blake would most vociferously object, it would be, arguably, Alain Badiou, current philosopher *par excellence* of Appearances and Events. Badiou's theorem that the unitary Being of the cosmos should be understood by means of mathematics and set-theory would astonish Blake, and would seem to incarnate the Newtonian Urizen himself, with his calipers and his scheming, Maoist logic. This world view would be, for Blake, the final tyranny of Reason over the fecundity of Imagination (as it seems to be, in fact, in Badiou), of the Symbolic over the Imaginary, no matter the differences between Badiou's mathesis and Newton's - Badiou's universe of sheer multiplicities over and against Newton's vast singularity, for example, or perhaps more to the opposite point, Newton's alchemical cosmos and astro/logical emphasis alongside his astro/nomical legacy. Or is Badiou's ambitious enterprise, which tries to account for the multiplicities of the world by a mathesis that sees the Void of Number at the heart of mind not simply a turn to emptiness, what the Buddhists might call dharmakaya, the primal Null set? We see something of this empty tension in Badiou's discussion of the relationship of art (here, poetry) to philosophy: whereas the world can be known (provisionally) through mathesis, there is no necessary relationship of art to the world per se. There is thus a break between the philosophical project and art, a break that engenders, I would argue, 'something else', something Other.

In his essay 'Language, Thought, Poetry,' for example, a seemingly purposeful allusion to Heidegger's book of similar title,[9] Badiou suggests that the poem (by which he means the lyric) is 'silent music', self-enclosed and non-referential, and so not any sort of representation of the world or experience. While he at first might seem to be invoking the epigones of formalism - -New Critical insistence on the self-sufficiency of the text, or positivism's insistence on the equivalence between language and truth to the exclusion of metaphor and ambiguity, or structuralism's investment in the relational import of information - this is too simplistic. Rather, in Badiou the poem, like any work of art, operates only and always within its own universe of correspondences, and so creates its own logic of event and appearances. Contra formalism, this universe of correspondences may or may not interact with larger social or political issues, but this is never of central import to the functioning of the poem. Badiou's larger point, in fact, is that the lyric departs from philosophy not because of its self-enclosed referentiality, but because of its non-discursive nature, its internal lack of argument about the actual truth of the world, hence poetry's 'silent music', a music that must (like music itself) remain mute about the condition of the world. But this, like Cage's Silence, seems less a silence than a silencing of aesthetic possibilities.

Indeed, 'silent music' also invokes, contrary to Badiou's intent, something quite Other: the long traditions of visionary and mystical tradition -

[9] Heidegger, Martin (2001) *Poetry, Language, Thought*, New York: Harper's.

St. John of the Cross uses the phrase 'silent music' to describe the state of mystical unity, while Hazrat Inayat Khan, the Sufi mystic, understands all music to be, first of all, silence. Unlike Cage, however, this silence is not simply a space to be filled with ambient sound. Rather, this silence is a creative potential that contains within it all sound, all music. For Khan, and the traditions he represents, music in fact emerges from silence, and is not, as conceived in the West (by, for example, Jacques Attalli among others), the product of noise organization. This silence is never an absence, never a lack, but is instead what Deleuze would call a virtuality, or what Khan would call 'the formless God', the uncreated creative impulse of the cosmos. So although this silence emerges in resonance with Cage's own understanding of silence as an impossibility, here the impossibility is an impossibility of pure possibilities, and not an absence, not even an absence which allows possibility to be contained within it (Cage). Silence, in Khan's thought, *is* virtuality, is *becoming* itself.

Badiou engages with Deleuzean becomings - and out of becomings, appearances - at various points in his work, most notably for this article, I think, in his single essay, 'Being and Appearance'. 'Appearance,' writes Badiou, 'is nothing but the logic of a situation, which is always, in its being, *this* situation'. Appearances, then, are particular - they cannot be totally abstracted, but must be apprehended in the here and now. This apprehension follows its own internal logic, which is precisely what allows appearances to be comprehended. From this he draws a somewhat Liebnizean conclusion (a conclusion which he later abnegates), '[L]ogic' he says, is 'the science of possible universes ... it is the principle of coherence which can be demanded for every existent *once it has appeared*' (Badiou 2004: 172; emphasis mine). The 'once it has appeared' is, I think, the sticking point, for here Badiou, though earlier assenting to the problem of becomings read through Hegelian dialectics, nonetheless engages appearances after the fact. In this passage (and it is not atypical of Badiou or cultural materialism in general) Badiou effectively erases the very space within which appearances arise. In focusing so closely on the 'existent once it has appeared' Badiou ignores becoming in favour of the already constructed. So, while giving credit to Badiou's attempts to account for becomings, for change, he is still, I think, a philosopher of outcomes (albeit outcomes 'read back' into their becomings) and final results. He does not, in the end, appreciate the radical possibilities inherent in process as process - in process that is essentially unknowable in terms of outcomes. The reason, moreover, that Badiou - and cultural materialism as a whole - privileges results and outcomes in this way is an inherent faith in materiality (and material practice) as the substratum of theory and praxis, a materiality that is, as I have suggested, being assailed on several different fronts at once.[10]

Badiou, then, like so many theorists (even, I would argue, Deleueze himself at certain points), negates becomings, and by erasing the potentialities inherent in becoming, implicitly posits becoming as a lack, and in doing so fails to consider that the seeming break or fracture between the artwork and the world, or between the artwork and philosophy, is a fracture of pure possibility, the space between being and becoming, the difference (and all difference, as Saussure reminds us, is silent) between the *thing* that is poetry or philosophy, and the becoming of the poem and the becoming of the world that philosophy describes - the difference between product and process. Badiou, as materialist thinker, cannot but apprehend *things* - including here appearances and events - and their logics within the context of their presumed completion, their 'being'. But the difference between things in their completed being and in their becomings is the difference between circumscription and possibilities not yet apprehended. This space, this fault-line is the trajectory of the visionary, the emptiness out of which new worlds appear and into which they disappear. The tectonic shift that opens this fault-line is the dance itself and

[10] 'Subject and object are only one. The barrier between them cannot be said to have broken down as a result of recent experience in the physical sciences, for this barrier does not exist.' Schroedinger, Erwin (1969) *What is Life? And Mind and Matter*, London: CUP, p. 137.

what appears out of and through this dance/ movement is an endless concatenation of becomings and disappearances, twining, twinning, one about the Other, sound following sound, movement generating movement, word generating word in an empty, generative grammar of infinite potential, a dance/movement that always eludes any final apprehension.

The gulf (or perhaps better, the *sound*) between Badiou's rather fixed understanding of the difference between things and the difference between poetry and *poiesis*, or philosophy and thought, or between being and becoming, is the space of *pharmakos*, expulsion, *katharsis*, that is both cure and dis-ease, exorcising the illusion of substance from the dream of materiality. When one attempts to 'read' the contours of appearances, or the internal logic of the event, one is 'always already' reading the given, a hypostasized and reified endless arrival already inscribed - and even though Badiou, acknowledging Deleuze, wishes to emphasize the flux, fluidity and shifting vitality of event and appearance, the object of analysis remains an object which is not merely readable but determinate. There is, then, always this remainder, an illusory materiality at the heart of mere appearance and event, an illusory materiality that wants expulsion and dissolution so that experiential emptiness, the substance of visionary truth - the truth of 'suchness' opposed to thingness - may appear. In such a world as this, the visionary is every bit as real as material practice - indeed, in a world such as this all realities are equal - all are equally real and equally haunted.

Blake created worlds, both possible and impossible. He insinuated alternate realities into our own, and spoke of multidimensional, coexistent spirit-realms. His eidetic visions - 'between sensation and image' - blurred the edges of what is meant by real, or even possible worlds and their virtual counterparts. In Badiou, for all of his seeming gestures toward becomings - becomings which end up as reifications - we are still in the realm of materiality whose logic of appearances and event always operates on the already-constructed, the about-to-be-constructed, or the projected image of some constructed totality. There is in this particular world and its logic of appearances seemingly no room for a visionary which rejects materiality not out of some misplaced Romantic nostalgia, as Badiou would have it, but rather out of the insight that materialism, as Gregory Bateson noted some forty years ago, is simply bad epistemology and false ontology. Indeed, if cosmology and quantum worlds have taught us anything at all it is that the substance of the world is by and large not substantive - the world is, if not nothing, very nearly so. Materialist philosophy has often shrugged its theoretical shoulders at this - what bearing does the emptiness of quantum reality or cosmology have on oppression, hunger, and exploitation that emerge from the material conditions, the material practices of culturally constructed economic and political realities? But this is to miss the point of visionary politics that sees in the fostering of the visionary itself the hope and the agency for political change - instead of relying on the already-constructed totalities of material thought, we might instead look toward what Deleuze and Guattari call the molecular, or what Gabriel Schwab calls the capillary rate of political change, the way that, as in Amazonian ayahuasca shamanism, the world arises and disappears in a continuous, moment-by-moment process, creating the world as it creates the curandera, all of it and her dissolving in the process itself. The ayahuasqueros took in, bit by molecular bit, the Maoi(st), N, N-DMT mixtures that transform thought, synapse by synapse, photon by photon. Ayahuasca transformed them, and they in turn transformed transformation itself into the very locus of power and resistance - feeding the medicine back to the white interlopers, ushering them into a madness and terror of their own making, politics and resistance becoming the visionary as becoming itself. Here, as in Taussig's tale of the devil-shaman, the logic of appearances is precisely a logic that eludes itself, erases itself, and transforms its own ground of becoming.

There are in the visionary no worlds given, no cosmos assumed, no materialities believed in - there is, or rather seems to be, simply the quantum foam of mind, giving supersymmetrical birth and annihilation to new planes of becoming, unimagined worlds giving birth to mind and emergent consciousness, imagining love and compassion, hard-edged and relentlessly empty, unencumbered by the nostalgias of the lost Romantic epigones of Euro American materialist phallacies. Here caduceus is the coiling image of divine mind, the messenger god, the sacred m-RNA. Ayahuasca becoming the vine of the dead and the serpentine dreams it induces, divine power and power of emesis - the theatrical moment of *katharsis* as revelation, as *anagnoresis*. Here snaky Hermes is not the fetish god of healing (though he is by now so associated in many minds) but of information as infection - the angel, the parasite, the artwork - patron of knowledge but also keeper of the hermetic, the hidden, the patron of the cheat, the con man and trickster, the actor as shaman - his caduceus coiling like the vine of souls itself, or like the Nigerian guinea-worm wrapped around the shaman's stick becoming, as the extraction is completed, a very symbol of its power, the power of the disease, the power of the cure, the power of a visionary making and unmaking of a world.

REFERENCES

Badiou, Alain (2004) *Theoretical writings*. ed. and trans., R. Brassier and A. Toscano, London, New York: Continuum.Brassier.

Bentley, G. E. (2001) *The Stranger from Paradise: a Biography of William Blake*, New Haven: Yale UP.

Shanon, Benny (2002) *The Antipodes of the Mind: Charting the Phenomenology of the Ayahuasca Experience*, Oxford: Oxford University Press.

Taussig, Michael (1987) *Shamanism, Colonialism, and the Wild Man: A Study in Terror and Healing*, Chicago: University of Chicago Press.

Tucker, Michael (1992) *Dreaming with open eyes: The shamanic spirit in Twentieth century art and culture*, San Francisco: Aquarian/Harper.

Weiskopf, Jimmy (2005) *Yajé, The New Purgatory: Encounters with Ayahuasca*, Bogota: Villegas editors.

Extracts from the Notebook of Xavier Valery

CARL LAVERY
in collaboration with
GERRY DAVIES

PRELUDE

Roughly five years ago on a typical bright April day in Paris, I found myself strolling past the charming stalls of the *bouquinistes* that hug the banks of the Seine, between the bridges that run from the Hôtel de Ville to the Ile de la Cité. Buoyed by the weather which seemed to herald the arrival of an early summer, I stopped at one of the stalls and leafed through the numerous books piled haphazardly on the rough wooden tables. I was immediately attracted by the appearance of an oversize text that had been printed by a publisher in Buenos Aires in the shape of an artist's notebook. Although the book was an attractive item in itself, elegantly designed and full of brilliant heteroclite images, I was seduced, above all, by the name of the author, Xavier Valery, whose surname seemed so close to my own that it could indeed have been an anagram. My interest was further quickened by the fact that the publication was in English (no translator was cited), and the whole feel of the book, both in terms of word and image, seemed saturated with a distinctly Anglo-Saxon melancholy. Even more mysteriously, it was impossible to determine if Valery was an author or, as I imagined him to be, a fictional character. And to this day, I have no idea if the book is a genuine autobiography or some form of postmodernist gag.

In what follows, I attempt to provide a reading of that strange text, entitled *The Notebook of Xavier Valery*, by focusing on Valery's response to the city. For what the author does so effectively is to make the invisible and affective life of the urban environment appear. In the pages below the reader will find my own attempts at exegesis and interpretation, as well as encountering examples of Valery's sketches, writings and drawings in charcoal.

A BILDUNGSROMAN

The first thing to take into account is that the book obeys the rules and conventions of what theorists of the modern novel call the *Bildungsroman*, a genre of writing in which the protagonist goes through a painful process of disenchantment and despair after an initial period of optimism and hope. In the final sections of the Bildungsroman, there is often a moment of dialectical resolution in which the hero - let's call him Julien Sorel after Stendhal's famous protagonist in the novel *Le Rouge et Le Noir* - finally discovers 'true knowledge'. Doubtless this move from thesis to antithesis and then on to synthesis explains why Marx, Engels and Georg Lukács were so beholden to this form of novelistic writing, seeing in it a perfect model for disclosing the hidden truth of bourgeois society. But things, for all that, have changed; history is no longer what it was. And in the hands of Xavier Valery, the *Bildungsroman*, despite chartering a journey from hope to despair, offers neither reconciliation nor truth.

Performance Research 13(4), pp.93-99 © Taylor & Francis Ltd 2008
DOI: 10.1080/13528160902875671

On the contrary, it ends in total defeat and dejection, as Valery, exhausted by his attempts to rewrite the city *à la* Michel de Certeau, and overwhelmed by a massive emotional event, which he only ever alludes to in the most elliptical of ways, decides to leave the territory that shattered him.[1] The last entry in the notebook, dated Easter 2007, for instance, is a direct citation, both ironic and telling, from the final page of Franz Kafka's novel *The Trial*:

> But the hands of one of the partners were already at K.'s throat, while the other thrust the knife into his heart and turned it twice. With failing eyes K. could still see the two of them, cheek leaning against cheek, immediately before his face, watching the final act. 'Like' a dog, he said: it was as if he meant the shame of it to outlive him (2000: 178).

If the narrative content or trajectory of the notebook charts the failure of Xavier Valery to rewrite the city psychogeographically, the form of the book is made up of fragments, drawings, sketches, aphorisms, philosophical musings and lists. In many ways, this heterogeneous assemblage appears to be borrowed from the US poet Charles Olson's notion of the open field, a poetics of energy charged with registering kaleidoscopic perception, the raw nerves of the subject, as he or she exists, bodily, in the sensorium of the world. The technical term for this writing of the body in poetic circles is 'proprioception' (Olsen, 1997: 181-90).

As I see it, Valery's notebook offers us a unique insight into how the city is lived and experienced by a subject who, more than most, was aware of what was at stake in the field of contemporary urban planning. Indeed, Valery's failure - his existential and sensorial defeat - allows us to see, through its very negativity, the need to create a new relationship to the built environment in which the senses are not only catered for, but actively celebrated. Arguably, the most depressing aspect of Valery's notebook is the profound sense of erotic misery that appears to have accompanied him for so long on his obsessive fugues across the zone. For him, the urbanised body is a body in pain, and I imagine him walking in the relentless rain, he so often evokes, like a modern-day Antonin Artaud or Alberto Giacometti, prey to a cruelty that he

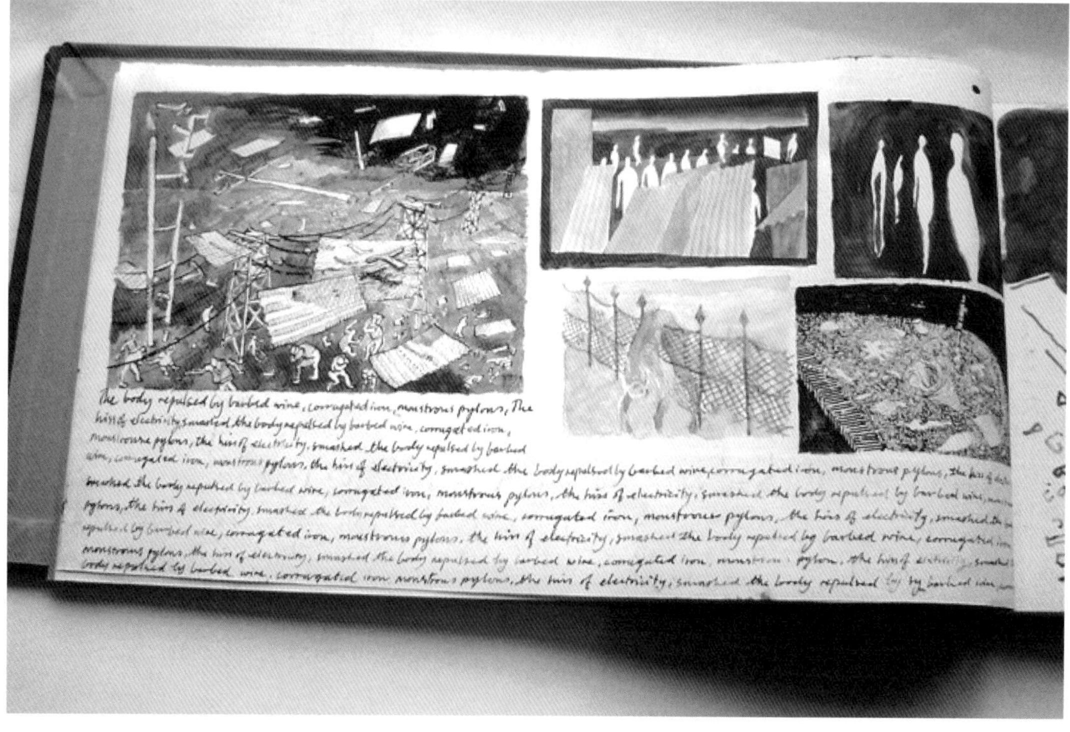

[1] I am thinking in particular of Michel de Certeau's overly cited essay 'Walking in the City' (1988: 91-110).

couldn't shake, in spite of his best efforts. Commenting retrospectively on the actions of Meursault, the character of his famous novel *The Outsider*, Albert Camus commented that in today's world Meursault is 'the only Christ we deserve' (2000: 119). I would like to borrow Camus's hyperbolic statement and to extend it to the sensorial peregrinations of Xavier Valery, a subject or character who sacrificed himself to reveal the sensual torture that inscribes itself, violently and invisibly, on the bodies of the numerous *Wandersmänner* who stride (anti-heroically) through today's cities in a world order on the very brink of collapse.

Extract from p. 112 of *The Notebook*

Date: All Souls Night
Title: On Cars, weather and iron in the soul
Epigraph: *'Pluviose irritated against the whole city'* (Baudelaire)

I can't help thinking how naive Jonathan Richman's song 'Roadrunner' is. Or then again maybe it isn't.

In it, he sings of being in love with the modern world as he drives through the suburbs of Boston on the highway all sparkling at night with the neon lights of electricity blinding him and filling his body with something ecstatic.

Richman is ethereal, all light, all speed ... an addiction to dromology, to not having a body, to being released from the body and being transformed into a flow of pure passing. Car = cinema. You look out at the world through a screen. And the screen protects you and nothing really bad can happen. I mean even when you crash; you don't

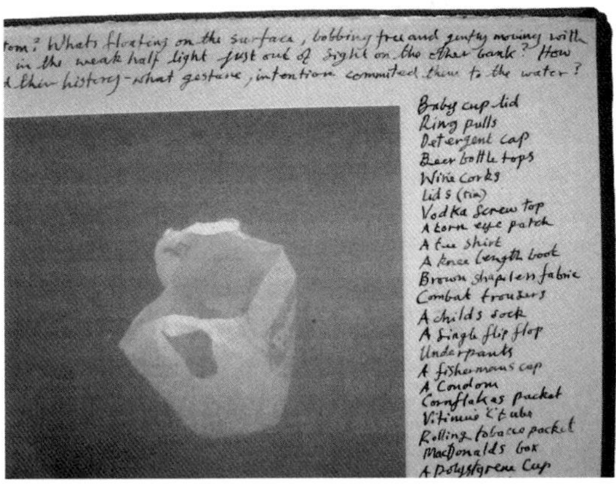

feel it. The whole world opens up like a big dream hole and the mind goes woozy as the world topples into an emptiness that swallows consciousness and opiates you just before you drop off into the soft crunch of metal on metal and forget that you've even been here at all. The car's medium is its message. Like Le Corbusier's machine for living made into a mobile metal box and separated off from everything. I mean vision dissolves when you look at the road in a car. In a car vision only really functions when you are at a distance, and the whole city, like in Richman's song, becomes hyperreal, a kind of film set, a simulation. The car is like a balcony; you are in the world, but not in it. Like dry masturbation.

But then again maybe not. Maybe a kind of pornography ... that's better ... more precise ... like driving through a sink estate ... and drowning and coming up untouched. You have registered everything. I mean it's all just information ... yes ... information ... little bytes for the eye ... and all the people there walking through the broken playgrounds in the mindless drizzle ... looking fat and forlorn and walking dogs ... .and not being able to afford petrol ... and boarded up pubs ... and wire ... .and pit bulls ... wolves ... horrible cross-breeds ... and heaps of dog shit melting and smeared in the drizzle on the pavement ... and then the mindless cars spraying up the frothy brown puddles that cling to the curbsides and everything is greasy and wet. Rain dirty. And then the night falls and there's no one there ... and the lights hit the streets ... and at moments like this I sometimes feel so sad, lonely and lost ... as if all my senses had just shut down in an automatic gesture of pure protection.

And I keep thinking of Roadrunner, and of how Richman is in the car, and of how my own experience of walking in the Zone is the opposite of all that ... ethereality ... I mean I'm all body, open to the elements and registering the vacuous emptiness of lots in the rain through my skin. And how that is pure NEGATIVE ECSTACY ... and knowing in an instant ... My time in the here and know. And realizing in a lightning flash that the body has been abandoned to all of this ... all of this what? ... I can't speak it ... I can't ART-I-CU-LATE. No capacity.

Again it is raining ... raining ... raining ... it's always raining ... HERE . I can't tell day from night now ... It's all so low ... the big rivers are swollen ... it's going to flood the ring-road ... and my body is damp and soiled ... my nose doesn't nose and my eyes don't eye. I'm all touch ... soaked through ... I can feel it in my feet and moving down through my shoulders ... and I'm shivering now and starting to get cold ... and I want to go inside. These bones ... bones ... my bones. There's a sort of shanty town at the back of the Holiday Inn where the bums sleep at night. I mean how can they sleep there ... they must have bad dreams ... all that stagnant water ... it stinks of shit here in the zone ... I mean it really does ... human excrement ... faeces ... dejecta ... waste ... The bums have found wheelie bins and put black plastic bags on the wet trees that stand alone in the bulrushes ... but it still stinks ... The whole Zone is full of little plastic bags of shit ...

WRITING THE BODY/ WRITING THE URBAN

Valery's writings register the immediate affect of the urban landscape on the body of the subject. What is undeniable in this body of writing - and the word 'body' is crucial here - is that Valery is not, in any way, recollecting his thoughts in tranquility (as the Romantic poet Coleridge once famously recommended). The writing here is intense, quick, marked by speed and rhythm. The writer appears flayed alive, as if his very skin were not enough to protect him from the sensory overload afflicting him. Valery's writing is nervous writing in the double sense of 'nervous'. Not only do his perceptions refuse to sit still, but they appear to come from the nerve ends themselves. Here, writing is a method of discharge, a device, a mechanism, for getting rid of experience. Hs words, or rather his sentence structure, exhibit a desire to get out of the world. As if he were shouting 'Anywhere but here.' The ellipsis or gap that Valery uses to join the fragments of perception and experience which he seemingly jots down at random is paradoxically the only thing that holds him, body and thought, together. There is no reconciliation or fusion in this writing. Tellingly, it is (often) devoid of

metaphor or simile, tropes that would bind him to something else, something bigger. This is writing as defense, writing as ballast. And as I end this section, I wonder if language, its syntax, grammar, rhetorical tropes, and figures of speech - that is to say everything that makes up a writer's style - might not be the best way to trace sensory urbanism in literature. Could it be that every city at a particular historical moment might elicit or impose its own objective style of writing, seek out a subject able to give a voice to its sensory reality? If this admittedly bizarre idea is accepted, we could then talk of a new way of reading the city in literature, in which content gives way to form, meaning to rhythm, mimesis to mood, description to musicality. Joyce's Dublin, Proust's Paris, Pessoa's Lisbon then could all be read as sensory registers of place, frontline reports for, and historical documents of, the affective life of cities.

DRAWING THE SENSES

At this point in my essay, I want to move away from writing as a system for representing the senses, and to concentrate, instead, on what is undeniably the most striking feature, if not the very *raison d'être*, of the notebook: namely Valery's drawings in charcoal. The drawings I want to show here are taken again from the middle section of the notebook and belong to a sequence named 'Crowds and Power'. Digitilisation means, of course, that some of the detail in the original drawing is lost, but I believe, nevertheless, that they still retain much of their auratic power.

At first glance, Valery's drawings in charcoal appear to exist in a contradictory relationship with the writing and the sketches. Where, in Valery's writings and sketches, the onus is on making sensory fragmentation as immediate and raw as possible, the drawings are, for obvious reasons, more composed and visually thought through. There is evidence here of reflection, stillness, balance. We can see a sophisticated aesthetic consciousness at work, filtering experience and rendering it visible. Nevertheless, the more that one looks at these drawings, structured, as they are, around a dynamic interplay between light/dark shadings, strong/weak lines, high/low angles and solid/amorphous shapes, the more the viewer is plunged, once again, into the maelstrom of painful affect. As with the writing, then, these drawings depict a body in pain, a body subjected to a great violence, a body assassinated by the criminal debris of a post- industrial world without alibi or reason, a body, in short, alienated.

Although I'm tempted to use the title of Richard Burton's book *The Anatomy of Melancholy* to describe these drawings, it is probably more accurate to see them as practicing an anatomy of abandonment. In them, we are presented, time and time again, with a body that has been robbed of its shape: ejected, dejected, and denied entry. Note for instance the amorphous homogeneity of the smudged masses who appear so desperate to enter the solid buildings of light, upon whose roofs nature exists as a kind of broken remnant, a tangled mass of leafless trees. This is a world of contorted shapes, blasted landscapes and utter despair, a gloomy world of surveillance and pain, a world of cowed heads and stooped shoulders, a world of cancerous absurdity, a world, that is, of total control. In these drawings, then, we are, I think, in the presence of something malignant, from which all hope of hospitality and welcome has been withdrawn, gone forever. That the drawings are made of charcoal, a natural substance, now burnt only serves to underscore the tragic irony of Valery's acute vision.

For me, and this relates back to my thoughts on Valery's writings, the drawings do not point forward to some future dystopia; on the contrary, they are rooted in the present, a distillation of contemporary urban experience. In this way, we can see the drawings as Valery's attempt to give substance to his perception, to bear witness in lead to the cruelty of the built environment in what I certainly take to be the post-industrial landscapes of northern England (even though the

book was published in Argentina). What the drawings communicate so well, so intensely, is a palpable expression of sensory deprivation. I do not understand deprivation here to signify an absence or lack, but rather to depict a form of bodily torture, an unbearable excess of what Lacanian psychoanalysis might call jouissance or painful affect.

I'd like to finish this interpretation by suggesting that, for me, the drawings of Valery disclose an intimate relationship between the senses and emotion. In other words, to exist as a sensory being is to exist as an emotional body. This reconfigures, I think, the whole notion of abandonment. Instead of reading abandonment ontologically or psychoanalytically as an existential event grounded in the subject's own private and phantasmatic relationship with the world, most notably, for instance, with their mother or with death itself, we need, rather, to think abandonment in terms of space, architecture and history, to understand its pain as being generated by the quality of our sensory existence in the world. In this model, if we can call it that, the architect replaces the 'shrink' as the physician for what is undeniably a sick world, a world in which the body suffers and where the senses have been jettisoned, or, more accurately still, subjected to a kind of complacent torture. Valery's drawings testify to what the land artist Robert Smithson might define as an 'abandoned future', a landscape of melancholic loss in which all hopes have been dashed (Smithson, 1996:72). However, there is ultimately more to them than that. Unlike Smithson, Valery shows us a world of betrayal, a world where justice has been perverted and where the architects of the present have designed, consciously, a world without a future. What I am struggling to say is that, in Valery's drawings, sensory deprivation is always political deprivation which explains why the masses in some of the images appear, like the Republican detainees held by the British government in the Maze prison in the 1980s, to be throwing excrement at the luminous archetypal building blocks that so dominate the landscape.

The author would like to extend his deepest thanks to Gerry Davies for 'sourcing' and working on the images.

REFERENCES

Artaud, Antonin (1993), *The Theatre and Its Double*, trans. Victor Corti, Calder: London.

Burton, Richard (2001) *The Anatomy of Melancholy*, new edn., New York: New York Review of Books.

Camus, Albert (2000), *The Outsider*, trans. Joseph Laredo, London: Penguin.

de Certeau, Michel (1988), *The Practice of Everyday Life*, trans. Steven Rendall, Berkeley: University of California Press.

Kafka, Franz (2000), *The Trial*, trans. Idris Parry, London: Penguin.

Olsen, Charles (1997), *Collected Prose*, Berkeley: University of California Press.

Rancière, Jacques (2003), *Short Voyages to The Land of The People*, trans. James B. Swenson, Stanford: Stanford University Press.

Smithson, Robert (1996), *Robert Smithson: The Collected Writings*, Berkeley: University of California Press.

Valery, Xavier (1993), *The Notebook of Xavier Valery*, Palmero Viejo: Buenos Aires.

Ambivalent apparitions: the pop-psychic art of TV medium John Edward

BRYONI TREZISE

THE POP ICON OF APPEARANCE: MEET JOHN EDWARD

Sydney's Entertainment Centre holds almost 13,000 people. Add to those people their dead relatives. Add again a few coincidental connections to the dead friends of those dead relatives, and combining dead and alive, psychic medium John Edward has an enormous crowd to please. I have been watching Edward's television programme *Crossing Over with John Edward* (secretly, addictively, at 3am) on cable TV for over a year now. But now Edward is here in person. His website claims that this Sydney tour sold out in fifteen minutes, placing him in the vendor's history book of bestselling acts of all time. A murmur erupts as the stage compère arrives to rev us up. We have come to see John Edward in action. *Clap.* We have come to witness the real thing. *Clap.* We have come to be in the presence of dead people. *Clap.* We have even come to talk to a few of our own. *Clap ... clap ... clap ...* and with that, John Edward - the world's first self-proclaimed 'media medium' - enters the arena (Edward 2001: xv).

As a dually live and televisual psychic event, *Crossing Over* offers a contemporary rendition of a traditionally arcane practice. Culturally, psychics have the job of enabling that which nobody else can master: making the disappeared reappear. In the west, this most commonly happens in the occult fairs and carnivalesque sideshows that house the religiously taboo. In this article, I examine how *Crossing Over* repositions the practice of ghost mediumship - a performance lineage most often aligned with premodern mystic arts - alongside the equally spectral modalities of technological postmodernity. In this, I read the exceeding popularity of *Crossing Over* as a much larger sign of the uncanny return of the mystical construct of appearance itself. It is in the very specificity of this tele-pop-cultural form that a highly politicised spectre of appearance can be perceived. This kind of appearance is predicated on a conflict situated within the term's own double entendre, where 'apparition' inscribes both an act of becoming (appearing) and a state of being (or seeming to be). I argue that *Crossing Over* strikes an oscillation between these two formal codes, using a postmodern mechanics of appearance to stage a premodern vista of appearance. Through this, the programme's central dramaturgical mechanism disrupts the very ideological effect that the programme attempts to maintain.

John Edward has joked that *Crossing Over* resembles a version of 'Opie does Dead People' - a curious blend of reality TV vernacular, game and talk show genres with paranormal content to boot (Edward 2001: 159). Tabloid writers have dubbed the programmme 'a monstrous hybrid of talk show and psychic fair, spliced with the look and sound of a 900-line infomercial' (Christopher 2000: 9), pitting Edward as a 'tomb reader', 'medium rare' and 'hustler' of the bereaved who exploits emotions for commercial gain. *Crossing Over* hence offers an exceedingly complex play with the tropes of trauma, selfhood and grief that proliferate as consumables within

Performance Research 13(4), pp.100-110 © Taylor & Francis Ltd 2008
DOI: 10.1080/13528160902875689

new millennial life. While its ghosts may indeed be fake, what the ghosts do is stage real memory effects - effects that encompass a rising social commerce between victimhood and visibility. This formal hybridity in enabling the dead to reappear positions *Crossing Over* as a hallmark cultural practice of late postmodernity - one that importantly conjoins 'old' occult technologies, 'new' televisual technologies and trauma discourse. While these dramaturgies work to mechanise appearance, they also reveal the performativities by which such appearance is enabled - namely, that what appears (the psychic, the narrative, the ghost) occurs through a collectively built erasure, compressing complexities of difference and identity into a suspiciously monocultural ghost-world.

In Edward's world, the figure of the ghost arrives not as a symptom of cultural grief, but rather as a figure that *assumes* the guise of loss and mourning to instead enact appearance's attendant ideological underpinnings - a proto-capitalist, neo-liberal politics built into a precisely staged apparatus for maintaining the normativity of social selves. These kinds of apparitions take the form of what Avery Gordon calls a 'seething presence', emerging rather as 'that which appears not to be there' but is (Gordon 1997: 8). This central notion of absence as a masquerade, of a ghost which *performs absence* places *Crossing Over*'s cultural haunt less in how Edward conjures the disappeared and more in how he erases the present. Like any good magician, Edward's game never appears to be what it really is.

THE CURIOUS INCIDENT OF A LETTER DISGUISED AS A GHOST

Edward (continuing on the subject of Dano's grandmother): "I'm getting 'Bo-bo.'"
Dano (perplexed): "Bo-bo?"
Edward: "Like two b's. 'B-b.'" (Dano is not making connection with any human she knows.) "Wouldn't be a dog, would it?"
Dano: "Beebee?"
Edward: "Passed?"
Dano: "Yes!" (Dano pauses and a look of amazement crosses her face.) "No! I had a dog named 'Beebee'!"
Edward: "Passed?"
Dano: "You get dogs?" (excerpt quoted in Christopher 2000: 9)

John Edward *gets* dogs. In fact, Edward claims that he can communicate not only with dead people and with dead animals, but with dead famous people, dead historical people and even with dead babies. 'What if we could have a snapshot of all the people John Edward said were clamoring to send a message to the living?' asks sceptic John Hockenberry. 'Well, they're not angry folks. They're not bitter at being taken too soon. We heard no stories of pain and suffering, no calls for revenge or settling scores. These dead people are nice. And maybe a little boring' (Hockenberry in Edward 2001: 243; emphasis in original). Painted in the picture of a picket-fenced North American suburbia (speaking to a predominantly Catholic Italian demographic drawn from Edward's own cultural background), even though his dead subjects have died gruesome deaths, fought family feuds or suffered grave and catastrophic accidents, once in heaven, at least according to Edward, there is peace, tranquillity and space for everyone - including the pet family dog. While part of Edward's psychic platform is the extraordinary ability to connect with non-verbal entities such as a dog, the image built around this particular skill not only confirms the superiority of Edward's psychic capacity, but stabilises the picture of a secure and Anglophone pet heaven (within a secure human afterlife) as the form of cultural imaginary Edward works to create.

Historically, ghosts are cultural disturbers. As the spectre of an often gravely unhappy end, the ghost is a symbol of narratives of morality and revenge, doomful fears about God and the hereafter and haunts that plague the present with fits of conscience or grief. This kind of ghostliness is disturbing for how it breaks the rules of sentience and cessation - ghosts are neither here nor there, and it is through this that they are truly otherworldly. As cultural actors, ghosts perform the act of becoming visible and likewise embody that process in ontological form. In this sense, their rendition of 'pastness', whilst seemingly engaged in looking back, is an effect of present concerns: they represent current preoccupations with the very matter of appearance itself. In his notion of the 'Ghost-idea' John Potts explains that different cultural contexts produce the terms of a ghost's conception and hence the capacity for its resulting sociological effect (Potts 2006: 13). Within the context of a post-private, digital age, Potts further suggests that mediatised technologies actually subsume the category of the ghost altogether. In this, the ontology of 'liveness', on which the ghost-idea conventionally relies, has been made redundant by the core discursive repertoire of the digital field. The ghost of technology evokes a specific kind of spectrality which effectively erases the authentic-mystic paradigm's traditional capacity to haunt. Ghosts no longer exist because their paradigms of disturbance (sentience and cessation) are no longer material concerns of the mediatised sphere. Haunting, it appears, only happens by way of informatic data flows or cinematic effect. Edward's work makes clear that in such a world of simulation, the appearance of appearance takes on a markedly different form.

The technological displacement of the ghost can be plotted in the drives for disembodied communication witnessed by the dawning industry of modernity and its correlative paranormal obsessions and effects. In early spirit photography, crossovers between medium and media were made theatrically visible by practices that tricked the light-sensitivity of the form. In these historical images, transparent bodies would float above the stiff poses of mid-nineteenth century life: vague ghosts, sad-looking ghosts, ghosts as skeletons and ghosts as ectoplasmic glue all flank the photograph's truest ghosts - the posing subjects themselves. Here families, portraits and the strange, stiff postures that arrive at a historical turn-point reveal the correlation between the drive of technology to capture life and the drive of mysticism to undo oppositions between body and spirit, life and death. Such drives were also made manifest in the late 1800s' paranormal complement to the arrival of the telegraph - the rise of spiritualism, a cult communication with the dead (Sconce 2000). Identified as the first technology enabling the flow of disembodied data, the telegraph importantly broadcast messages across wire, a revolution in the bodily ownership of the word. Spiritualism interestingly channelled the deceased through imitative rituals which used automatic writing or séance to counter - but also meet - the new sciences of modernity. As Erik Davis notes:

> While the technology of the telegraph transformed America into a wired nation, the *concept* of telegraphy enabled endless displacements of agency, projecting utopian possibilities onto a disembodied, invisible community and recasting an often radical political agenda as an act of supernatural possession. (1998: 49)

As these tandem histories suggest, it is in the crossovers between digital and divine, and their implication in the construction of a broader public imaginary, that an emergent politics surrounding the very matter of appearance can be perceived. The ghosts of modernity were not merely remnants of the past, but effigies ghosting variable crises in the practices of self and polis amidst dense social and industrial change. That the conceit of mobilised agency signalled by early telegraphy was in fact symptomatised by Spiritualist telepathy hence prompts a reading of the contemporary televisual apparatus as not only beset with spectral qualities itself, but with a similarly political agenda imitated in the ghosts that it constructs. That is, John Edward's ghosts recast the politics embedded in the idea of televisual agency through their enactment of a distinctive mode of appearance, one which takes place consecutively in both live and televisual form. And yet, Edward's imaginary - a stabilised pet-friendly ghost-heaven - occurs at a pivotal cultural moment that evokes different discursive refractions from those of mere moral warning or ancestral respect. Edward's ghosts work against their own ontological displacement by appearing at the very time that witnesses the erasure of the ghost altogether. As distinctly postmodern ghosts - televised reappearances of the televisually repressed - Edward's ghosts are a curious cultural paradox: they discursively re-appear reappearance itself in the image of all things, of Beebee the ghost dog.

CROSSING OVER: THE RULES OF THE GAME

Crossing Over is a televisual transmission of a live performance that occurs between a medium and their client - a practice that Edward claims he's 'been doing for years ... with just God's help' (Edward 2001: xx). As a live event, Edward offers memory traces that link the alleged spirits of deceased relatives to audience members via a process of data validation, usually given by the calling of information that he 'gets' from 'above'. He firstly receives a message in the form of a letter trace - a sound that registers as 'two b's', a 'b-b' or a 'bo-bo'. As a first step in the programme, Edward locates the area within his audience that the clue might be aimed towards and then produces the clue as data that he has been sent by the ghosts themselves. Clues usually begin with letters that signify names, but can also be in the form of numbers interpreted as special dates: a birthday or anniversary, for example. Other clues can involve the description of a death, an image of a personal item, or a spoken phrase. An audience member's role within Edward's production of data traces is to validate those traces. Validation ideally occurs with simple negative or affirmative responses, but usually occurs with segments of elaborated information that bolster the otherwise uncontextualised trace Edward has provided.

As an initial step in the execution of Edward's live mediumship, the trace-giving process is the first appearance amongst a chain of appearances entailed by the *Crossing Over* format. This first stage is a process that involves the whole audience in the reception of a ghost and in the validation of the ghost-world. The pause in the excerpt above, between Dano's confusion over 'b', and her later acknowledgement of 'b' as 'Beebee', indicates how the central ambivalence of the ghost traces work as a suspended referential leap during which the entirety of the audience can claim the memory signifier Edward emits. This moment is essential to understanding how the trace generates memory as *affect* - an experience of memory that lacks the content of memory at its base but is nonetheless experienced as the embodied authenticity of memory as social discourse. For the duration of the time that the trace remains unclaimed, the audience filters their own cognitive data for a match. It is here that projecting onto a trace provokes the possibility for the trace to become story. As a collective force, the audience supports the referential effect, or deictic power of the trace through shared desire. As narrative builds into the trace, Edward feeds in more traces to locate a singular audience member as connected to the ghost and to expand the moment of mediumship into a moment of confessional story.

The audience in Edward's live gallery negotiate positions of viewer and participant. They witness readings conducted for other audience members and anticipate readings that may be conducted for themselves. While these dual positions require different techniques of performance and reception, they also place viewers in differing subject positions with regards to the economy of the programme. During the course of a reading, a viewer can expect to narrate, guess, watch or confess, for example. This is not simply a signposting of differing emotional responses, but is much more significantly a process of playing out relationships to the cultural commerce of appearance itself. While each modality speaks to variants of the ideology of appearance, it is the *shifting between* these modalities - or more importantly, how this shifting makes these modalities distinctly ambivalent, or even defunct - that generates the premise for a postmodern ghost practice. Where successful appearance signals the correlation between participant subjectivity and televisual visibility, Edward engages a rehearsal of subjectivity built out of consuming the collectively, imaginatively traumatised. As a memory text based on reproducing the traumatic past for the present, the very tenets of appearance become implicated in the kinds of cultural commerce Edward sets up.

The semblance of appearance becomes apparent in the oscillation the programme strikes between a stable ghost-world on the one hand, and the highly unstable process by which it establishes this ghost-world, on the other. This occurs in how Beebee the ghost dog signifies both a *thing that is named* and *a linguistic repetition* that is sonic but not meaningful in itself. In being both a sound pattern and a lost object, Beebee signifies signification: she is both a random sound-pattern produced to make meaning, and simultaneously, the resultant memory-effect of a referential sequence. As a cultural memory trace, Beebee hence collapses both the process of enabling appearance (a linguistic game) and the desired result of appearance (a ghost) within her singular but complex ontology. In being both a name and a phonetic letter, Beebee essentially alphabetises herself into being. Thus 'b' becomes 'Beebee' becomes a lost family dog. As an audience waits for Edward to confirm that 'b' is really Beebee, and as they then wait for Dano to confirm that yes, there was 'a dog named Beebee', *Crossing Over* articulates its ideology of appearance as the coincidence between a highly arbitrary practice of naming and a highly traumatic practice of mourning or dying. Performatively speaking, Beebee is *both* the process of naming and the culturally embedded after-effect that the naming produces: a phoneme cleverly *disguised as a ghost.* What Edward is playing at, as much as memories, is the semiotic system: the game of meaning-making itself.

Beebee characterises how the ambiguous terms of the *Crossing Over* game produce the apposite terms for the secure social imaginary of the ghost-world. In this sense, the boringness that is central to Edward's supernatural, typified in the image of a dead family dog, can be actually understood to signify *a world that ghosts a much larger crisis* - not a crisis of death nor accident, *per se*, but a rather a crisis of reference. This larger crisis can be recognised as working into the very structures that build everyday subjectivities and practices of faith in concrete action and knowledge. In this, Beebee is a motif revealing why *Crossing Over* needs to produce a highly formulaic ghost-world to hide (or re-ghost) the play of chance that it enacts across the socio-semiotic field. The plain mechanics by which Edward is able to make his live programme work, and also by which he is able to televise his re-edited programme to a bulk home-viewing audience, are hence made clear in the performativity of Beebee's ontology. The connection, however, between Beebee's cultural identity (her ghost effect) and her linguistic ontology (the game of appearance) is here pointedly marked. We begin to see the game-like logic that connects a semiotics of loss to a cultural thematic of loss. While the mechanics of appearance are essentially arbitrary, built around endless combinations of signs, these nonetheless stage real memory effects. While Edward's audience want contact with what they have lost, what they actually do is unwittingly enact the loss of loss itself.

MEDIUM RARE: CROSSING OVER AS TRAUMA TV

A good psychic needs a good back story to back them up. Edward not only has a syndicated television programme to his name, but conducts live seminars, private readings, has appeared on *Larry King Live* and has authored numerous videotapes and self-help books. His most recent book explains how he became what one reviewer describes as 'one of the few growth industries in an otherwise lacklustre economy' (Jaroff 2001: 52). Edward's story began as a 15 year old boy in moments where he 'knew things' that he 'shouldn't have known, family events that ... no one had told me about'. He refers to his spiritual guides - 'the boys' - as keepers of his sixth sense:

'They were leading me to the understanding that I was on a path to a life's work connecting the physical world to the spirit world' (Edward 2001: 5). He also describes his mother's death in his youth as pivotal - the reason that he converses with the dead for a living. From the development of psychic skills, to debuting at psychic conventions, to appearing on radio talkback, to writing a book that flopped in the market, Edward at least sees the irony in his media mediumship: 'I see dead people' is a catch phrase with which both he and Bruce Willis have been associated.

Edward is uncannily aware of his co-presence as a media icon and psychic medium, characterising his work as a 'blend of spiritualism and entrepreneurship' (Edward 2001: 7). His success is demonstrated in *Crossing Over*'s overnight ratings high, with *People Weekly* reporting that since premiering on the Sci-Fi Channel in 2000, it became the most popular programme in syndication with an audience of more than 3 million - making Edward into a 'whole other sort of astral projection: a TV star' (Gliatto and Stoynoff 2002: 85). Tabloid press are nonetheless tentative around his handling of delicate subject matter. The blending of spiritualism with the mercantile world presents a sticky ethical dilemma. Edward's critics have varyingly debunked *Crossing Over* as the ruse of eavesdropping, of planted actors and of cold-reading strategies - ploys in which a medium delivers a stock questionnaire to persuade an individual that he knows all about their problems. *Time Magazine* ostensibly uncovered Edward's scamming, accusing him of less than psychic ways: 'aides were scurrying about, striking up conversations and getting people to fill in cards with their name' which were 'picked up by the microphones strategically placed around the auditorium' (Jaroff 2001: 52). Edward's standard response to such allegations is to replace the terms of provability with morality: '[w]hat does get me upset is when people try to define my motivations ... Saying I'm in it for the money or that I'm taking advantage ... or that I'm a phony' (Edward 2001: 238). Set thickly amidst the tropes of televisual programming, the business of *Crossing Over* is in the question of what his ghosts have been exactly conjured to do.

Edward's gallery is intensely serious. The episode channels a teenage girl who died suddenly from encephalitis - viral swelling of the brain. Edward connects her with her grieving parents:

> John: "I'm supposed to acknowledge for you that there's surgery that happens in the head area, okay?"
> Parents (crying): "Yes."
> John: "I want you to know that I know that you're thinking that something went wrong, okay?"
> Mum (crying): "Yes."
> John: "I know that you're thinking that."
> Mum (crying): "Yes."
> John: "But you have to understand that the type of surgical procedure that was being undertaken was huge, okay? It feels huge to me and very, very delicate. And I feel like what happened with her was explained to you in some way. I don't know if this was like an emergency surgery, I don't know, but I feel like I need to stress to you, I feel like it's explained in detail: these are the precautions, these are the things, the negatives ... It feels like it's a freak thing that takes places within side the surgery itself; a reaction to the procedure, a reaction to the anaesthesia, there's an inflated feeling coming within side me so I don't know if a lobe of her brain started to swell, okay, but there's an inflation that they're trying to show me (parents nodding, crying). The reason why I'm being this graphic is because she's being this graphic, and the reason why she's being this graphic is to validate for you both - (to the father) I don't know if you're Dad - but to validate for you both that she is okay."
> (*Crossing Over* November 2005: Arena TV Australia)

I watch this episode on television. In the televised epilogue, the parents describe the psychic experience as the bittersweet quality of 'going to visit somebody that you haven't seen in a long time and knowing that you're never going to be able to see that person again'. The studio audience appear to be still, even stiff. It seems that even if they are not convinced that they are in the presence of an actual ghost, they know that they are in the presence of trauma. This, and the kind of story every person dreads to make their own.

Reality television has been noted for its easy assimilation of multiple generic forms, where a 'deep-seated and institutionalised blurring ... between performance, mediation, narrative and fact' has come to signify a new age in televisual practice - one in dawning since the 1980s (Dovey 2000: 10). Within these formats, the trope of authenticity drives a diverse range of surveillance, documentary, chat show, competition or docudrama genres that now characterise the excessive visibility enabled by the more generalised mediatised field. What is important about these multiple reality formats however, as John Dovey argues, is not only their shared predilection for making intimate revelation a 'key part of the public performance of identity', but the fact that each mode is decoded by viewers with an acute sensitivity 'to precisely [the] kind of reality reference any given show is based upon' (2000: 10-12). *Crossing Over*'s reality reference is firstly produced by the internal genre-referentiality the programme draws upon: it is a chat show, a psychic show, a trauma programme and a live audience competition, where each formula works to stabilise another format. The programme's unique reality reference is secondly built upon an interesting inversion, but utilisation, of the reality form. While it *appears* to be a programme that uses a reality TV format to paradoxically channel paranormal content, it is rather a programme that uses the pretence of the paranormal - *the apparatus of appearance* - to force a reality television result.

This kind of result can be thought of in terms of what Laura Grindstaff has theorised as the reality TV money shot (Grindstaff 2002). Drawing on theories of pornographic performance and spectatorship, Grindstaff argues that the money shot, more usually conceived of as the moment of sexual climax, can be characterised in reality TV as the moment when personal trauma becomes televisual spectacle - moments when seemingly ordinary people explode. These are understood as instances when studio guests lose control and express joy, sorrow, rage, or remorse on camera. Fans contend that what makes Edward uncanny is 'how he nails down concrete details from wispy hints' (Gliatto and Stoynoff 2002: 85), and how he can 'crumple audience members into hysterical blubbering by giving proof - through dates, names or remembered objects - that their dead relatives were saying hello' (Wolk 2001). As one cameraman's response to an episode makes clear:

> I was shooting the close-up and watching the expression on his face, watching him begin to fall apart. First the tick and then the trembling in the chin and the kind of embarrassment, the eyes shifting while seeing if anybody's watching. And I literally watched this person fall into this grief and surrender to it publicly. It was very hard to watch. (Edward 2000: 222)

Further to this, the live moment of confessional grief is bolstered by additional memorabilia. In the transcribed episode above, the programme visits the deceased child's parents and films the interior of her neatly preserved bedroom. Family video footage, photographs, interviews and objects all further co-construct the authenticity of the appearance. Face-to-face with the after-effects of death, a home viewer receives a picture of a life compiled through a now made public personal archive. Televisually, the deceased person becomes a reconstructed object, an ironically produced after-event in themselves.

Crossing Over uses the conventions of reality television to play at multiple truth regimes simultaneously: the authenticity of ghost ontology prefaces the authenticity of what Dovey terms 'trauma TV' - 'individual tragedies which would have once remained private but which are now restaged for public consumption' (Dovey 2000: 21). Edward works on the premise that he conjures ghosts in order to heal - in his words 'to validate for you both that she is okay'. From a discursive viewpoint, however, Edward's ghosts appear for their money shot capacity: they produce consumable trauma. More than this, in the varied stages of their ambivalent emergence, they construct a tension between live participants and televisual spectators as the trauma ownership is collectively bartered and negotiated. This tension is taken further when *Crossing Over* performs a televisual 'crossover' with other televised events. In such a complication, Edward devised a themed special to revisit his 'favourite' daytime soap opera, *The Guiding Light* (17-21 November, 2003) - itself based on a narrative of ghosts. In this, viewers could 'tune in all week as [their] favourite celebrities [actors from *The Guiding Light*] connect[ed] with their loved ones and share[d] incredible stories ... that out-soap the soaps' (*Crossing Over* November 2005, Arena TV Australia). Within this episode, Edward was to give psychic readings to the cast of *The Guiding Light*. The *Crossing Over* reality-reference hence deferred itself not to external worlds, but rather returned itself to the interiority of the televisual world. What emerges is a genre complication that produces *Crossing Over* as a 'real life' psychic soap opera and that doubly reinforces the paranormal fiction of *The Guiding Light*'s subsidiary text.

Dovey's analysis of reality TV formatting is interesting for the connection it makes between the reality genre and its correlative production of political economies, economies that ideologically reproduce the versions of normative citizenry that contemporary western sociality aims to maintain. In his argument, Dovey suggests that neo-liberal economics firstly disenfranchise resources for the making of more balanced or nuanced documentary texts at the expense of cheapened production values and a rationale of outputs over content. Secondly, however, they signal an insidious turn in the actual function of the televisual and its role in promoting a version of a unified global imaginary. He writes:

> The significance of Reality TV hybridisation lies not in the way it signifies either economic imperatives or postmodern genre collage but in the way it insists on the primacy of the individual, emotional, and above all unified version of subjectivity. (Dovey 2000: 97)

In this sense, while the televisual format of a cultural performance such as *Crossing Over* manipulates the very postmodern excesses of hybrid form, such extensive hyperbole actually disrupts the multimodal nature of the programme to instead promote versions of unity, truthfulness and authenticity given firstly in the ghost-world it sets up. To this degree, audience members unwittingly play at maintaining the redundancy of one system which is always already operating in the service of another (hidden) one. Rather than its ghosts, this is possibly the biggest ruse *Crossing Over* constructs as a televisual social sanctum.

Alongside the cultural demise of the ghost, Dovey argues that the burgeoning effects of technology are equally responsible for a demise in traditional documentary form: one originally invested in a democratic public sphere of 'truth' rather than in an electronic super-space based on an unprecedented inversion of public and private. In this sense, what was once a genre thought to structure a collective communicative space now structures the very ethos of that space in simulation: it maintains a pretence of the constitution of democratic play but nonetheless reinscribes beneath this regime a play of contradiction and ambivalence as the marker of its truer mechanics. For Dovey, this confirms what all reality television genres generate: 'the production of normative identities' which propose to *restore* 'lack of narrative coherence' so that '[s]ubjectivity, the personal, the intimate, becomes the only remaining response to a chaotic, senseless, out of control world in which the kind of objectivity demanded by grand narratives is no longer possible' (Dovey 2000: 26). Postmodern collage is hence co-opted to produce its opposite: a performance of unified subjectivity seen in grieving parents, an empty bedroom, a young girl's tragic death. Such images reveal how *Crossing Over* not only constitutes, but continually *reconstitutes*, a version of the self as it relates to publicly constituted knowledges and systems of power. The identities within the programme emerge not only as content to be consumed but as a practice of *performing consuming*. While its pretence is the appearance of ghosts, *Crossing Over*'s actual reality-effect works to *display subjects* in their continued and unwitting *performance of the attainment of subjectivity itself*.

The twin ideas of a particular reality reference (the production of an exceedingly simulated appearance) reproducing a particular regime of truth (the production of secure and stable knowing) emerge in *Crossing Over* in the way that participants perform themselves as distinctly fragmented against and in pursuit of that which is perceived to be whole. Enforcing the regime of truth, then, is a regime of power by which participants attempt to attain discursive control over a system that is in effect always already defunct. Discursively, we witness subjectivity unravel and be remade as participants work to generate a collective ghost pool, but rather refigure themselves as the primary ghosts of Edward's vision. In their failure to register as single and complete, Edward's audience rather pursue - in the image of a secure ghost-world - their own discursive deaths.

LOSING LOSS: THE OCCASION OF A 'ME-TOO'

As the Beebee example illustrated, Edward enables appearance to emerge in the relationship between a memory trace held by a viewer and the mnemonic effect generated by the calling of linguistic structures: a 'b' becomes 'Beebee' becomes a lost family dog. While the main task of *Crossing Over* is to secure this referential leap in order to produce a requisite money shot, Edward's role as convenor of this process is to manage the flow of memory traces between mnemonic triggers and audience identities - it could be argued that he in fact manages the cultural apparatus of appearance itself. This is a process which interestingly fails when audience members cannot align with the traces Edward provides. In situations such as this, the collective process of generating a ghost breaks down and the apparition acquires an undesirable liminality through the conflicting validations it summons. These are importantly instances in which the ghost, as a secured and stabilised idea of appearance, begins to disappear. It is no longer a ghost, it is not even that. As an uncanny showman, Edward's phrase for justifying these slippages is a 'me-too': moments in which multiple ghosts appear at once, confusing Edward as to who is what ghost, belonging to what family, speaking to what audience member. Discursively, this can be read as a moment in which varied traces and their respective audience recipients are in competition, leaving the errant memory signifier suspended without chance of referential closure until it is singularly claimed:

> John: "The first person I'm coming through to is a male figure to the side, a husband or brother." (To Andrea) "You're saying your brother is passed?"
> Andrea: "Yes."
> John: "I'm going to tell you that your brother, if it's him, is bringing through an older male, so your father, your father-in-law or there's another male figure above ... "
> (Andrea cannot acknowledge this)
> ...
> John: (Voice-over) "At this point it seemed these messages were for Andrea and her husband Dan. However, sitting right in front of them was another family and judging by their reactions, I could tell that Jean and her daughters wanted to join in. This reading was about to become a classic example of a 'me-too', where two or more different families have loved ones coming through together."
> John: (to Jean) "Are you connected to the Joe?"
> Jean: "John, my father's name is John."
> John: "Woah, hold on a second ... I have the male figure to the side who has to be through first, who wants me to acknowledge that the father figure is with him there. There is a J-o name like John, like Joe, I thought it was Joe. And there is a lung cancer, emphysema, blackness to the chest, connection and ... the third month, March, or the third of the month is connected to this in some way." (To Andrea) "Did your brother pass from a suicide?"
> Andrea: "My brother? No."
> John: "There's also someone who passed from a suicide where their actions brought about their passing. What they did caused how they passed."

> Andrea: "My friend was in a car accident."
> John: "No, no, no, this isn't a car accident. This is either somebody who either accidentally electrocuted themselves ... there's an accidental type of ... but they take the responsibility on for how they passed."
> Andrea: "I had a cousin."
> .... .
> John: "I'm not really sure who's coming through, or which family I'm talking to, to be quite honest. I don't know if it's your family or your family ... "
> (*Crossing Over* November 2005: Arena TV Australia)

In this excerpt, Edward first introduces Andrea's 'brother figure' who is also in the company of (Edward uses the term 'bringing through') a 'father figure'. Andrea cannot validate the clue of the father figure and so her ghost then becomes Jean's ghost and is, in this confusion, potentially *not Jean's ghost either* - it is strangely both or neither, and the entity which registers 'a J-o name' and a 'blackness to the chest' hovers above the room as an unknown spectre waiting to be claimed. And yet it is the behaviour (read: desire) of the audience members who attempt to match the data that keeps the possibility of appearance intact. Andrea's response to Edward's description of a suicidal death with the declaration that her 'friend was in a car accident' reveals the struggle for identification audience members push to experience. As television critic Josh Wolk observes of another reading:

> Edward was getting a K name who died from a bad blood transfusion. When no one spoke up ... I felt those around me struggling to remember a connection that could help them claim it. 'It's kind of like a psychodrama where there's a willing suspension of disbelief ... All the players are integral to it'. (Shermer in Wolk 2001)

The significance of a 'me-too' not only rests in how it reveals the building of an accepted ghost-world through collective profiling, but how it reveals what spectators practice of subjectivity as it is tied to ideologies of appearance within this process. While the aim is to reach a stable image - a fantasy ghost-world - the process of arriving at that image conflicts with the establishment of that world: a process based on performing reference in crisis. For Andrea, it is neither completely her brother nor father, nor not her brother nor father who communicates. Instead, the most direct communication exists between herself and her contender, Jean, as they try to affirm themselves through the ghost traces they are given. At the centre of this 'me-too' is the fraught proposition that for a participant, the stability of a subject position promised by the programme is continually undermined by a central ambivalence surrounding ghost ownership. This can also be understood as a competition for dominance of the social semiotic space. What Andrea and Jean want to lay claim to is the purchase of a single letter 'J' - a letter disguised as a ghost - and its attendant promise of discursive power. Instead, Jean and Andrea become witnesses to their own discursively built experience in the very moment that the phoneme 'J' will not materialise as a ghostly effect, but rather materialises their desire in play. When faced with losing the letter 'J' and its promise for generating a narrative of loss, what both Andrea and Jean experience is rather the loss of loss itself.

GHOSTED GHOSTS: THE MISSING EPISODES

I have so far argued that Edward's ghosts play the game of appearance by producing a memory negotiation that posits loss, as well as the loss of loss, as central to the construction of their ontology. Within this process, a spectator can remember another's memory, can remember things that never happened, and can play at being a social rememberer. It is in this respect that the stakes of losing loss become exceptionally high, especially when Edward's work transgresses its inter-televisual references to place itself amidst a global trauma referent: the 9/11 attacks on the United States of America. In this final example, introduced only by way of concluding, a 'meta' instance of losing loss is made apparent by themed readings Edward conducted with families of 9/11 victims shortly after the attacks. Of interest in this account is not the episodes themselves, but the fact that they were suppressed before airing: Edward's ghosts were made to vanish before they were even created.

As I explained above, it is the collapse of the ghost in a 'me-too' that forces participants to extend beyond their consumption of trauma and to recognise their complicity in making grief operate within an economy of visibility. The similar, but more politicised, loss of loss engendered by the suppression of the 9/11 episodes points to a parallel moment of perception: a consumptive audience is left bereft of the very master trauma signifier they both want to claim, and have produced. It is in this respect that the televisual apparatus of appearance, and the presence or absence of Edward's ghosts in that apparatus, holds tremendous stakes. As many of its commentators have noted, September 11 was enacted to damage the very core of the western world's representational apparatus. In doing so, it inverted the parameters by which representation and experience are tied, parameters whereby - in Slavov Zizek's terms - 'we lived in our reality, perceiving the Third World horrors ... as something which exists (for us) as a spectral apparition on the (TV) screen' (Zizek 2001).

Re-absenting *Crossing Over*'s September 11 ghosts from the framework of consumptive othering the programme generates, positions Edward's spectres as distinctly *different from* the interminably 'real' images of ethnic and middle eastern otherness the staged hypervisibility of the 9/11 attacks sought to cathect. In other words, the national suppression of Edward's episodes points to a dramaturgy of appearance that only enables certain kinds of bodies to appear as *appearances*. A moment of losing loss offers the possibility of re-perceiving appearance through a practice of ambivalent mourning that restricts the consumption of trauma to instead envision the self as it enmeshes with stories of, visions of, and ghosts of the 'other'. As made clear by the missing 9/11 episodes, the ghosts that really hide within Edward's simulacrum are the products of the hypervisible sphere of western hegemony itself - a scenic apparatus masked most lucidly by the image of a happy but dead family dog.

REFERENCES

Arena TV Australia (2005) *Crossing Over* (November) episodes transcribed by the author.

Christopher, Kevin (2000) '"I Speak to Dead People": Medium John Edward Hosts Sci-Fi Cable Show', *Skeptical Inquirer* 24.5: 9.

Davis, Erik *TechGnosis* (1998), New York: Three Rivers Press.

Dovey, John (2001) *Freakshow*, London: Pluto Press.

Edward, John (2001) *Crossing Over: The Stories Behind the Stories*, New York: Princess Books.

Gliatto, Tom and Stoynoff, Natasha (2002) 'Medium Rare' *People Weekly* 57.17 (6 May): 85-86.

Grindstaff, Laura (2002) *The Money Shot: Trash, Class and the Making of TV Talk Shows*, Chicago: University of Chicago Press.

Gordon, Avery F. (1997) *Ghostly Matters: Haunting and the Sociological Imagination*, Minneapolis and London: University of Minnesota Press.

Jaroff, Leon (2001) 'Talking to the Dead', *Time* 157.9 (March 5): 52.

Marshall Clark, Mary (2002) 'The September 11, 2001, Oral History Narrative and Memory Project: A First Report' *The Journal of American History* (September): 569-579.

Potts, John and Scheer, Edward (eds) (2006) *Technologies of Magic: A Cultural Study of Ghosts, Machines and the Uncanny*, Sydney: Power Publications.

Sconce, Jeffrey (2000) *Haunted Media:* Electronic Presence from Telegraphy to Television, Durham and London: Duke University Press.

Wolk, Josh (2001) 'Tomb Reader' *Entertainment Weekly* 614 (14 September): Webpaper http://www.ew.com/ew (accessed 25 June 2007).

Zizek, Slavov (2001) 'Welcome to the Desert of the Real'. Webpaper www.lacan.com/desertsym.htm (accessed November 2002).

Appearing to Play: A Memory Toy Theatre to Cut-Out and Collect

Performance Research 13(4), pp.48-55 © Taylor & Francis Ltd 2008
DOI: 10.1080/13528160902875697

The Galapagos Man, A Toy Theatre.

Tab 1.

Tab 2.

(Attach TAB 3)

(Attach TAB 4)

(Remove Centre Piece)

(Attach Foot TAB)

Control Rods for each standing character, S1–S6

H1–H13 are all hung characters/objects.

Scenario One: An Opening.

1. Turn on torch. Direct a long slow sweep back & forth across the stage – eventually resting in the centre.
2. Gradually lower DARWIN THE FINCH front stage right of fly bar one until positioned half way from bar to stage. TIE OFF.
3. Turn off torch. Turn desk lamp on.
4. Slowly move THE MOTHER APPARATUS from back stage right to left.
5. Turn desk lamp off. Turn torch on.
6. Gradually lower BIRDCAGE from back stage right from fly bar 2 until it sits at the same height as the table on the backdrop.
7. Lower Galapagos Man Puppet front stage left from fly bar 1.
8. Enter GALAPAGOS MAN TOPLESS stage left.
9. Hold still for 7 seconds.
10. Turn off Torch.

Lumber Room Set

- 6 Tables run across width from Table 1 stage left to table 6 stage right where the LUMBER ROOM DRUM is placed.
- Table 4 has a stove for cooking

Attach Tabs to Sides of Card.

(Attach TAB 1)

(Attach TAB 2)

3.

4.

The Lumber Room

Attaches to the foot of proscenium

Foot TAB

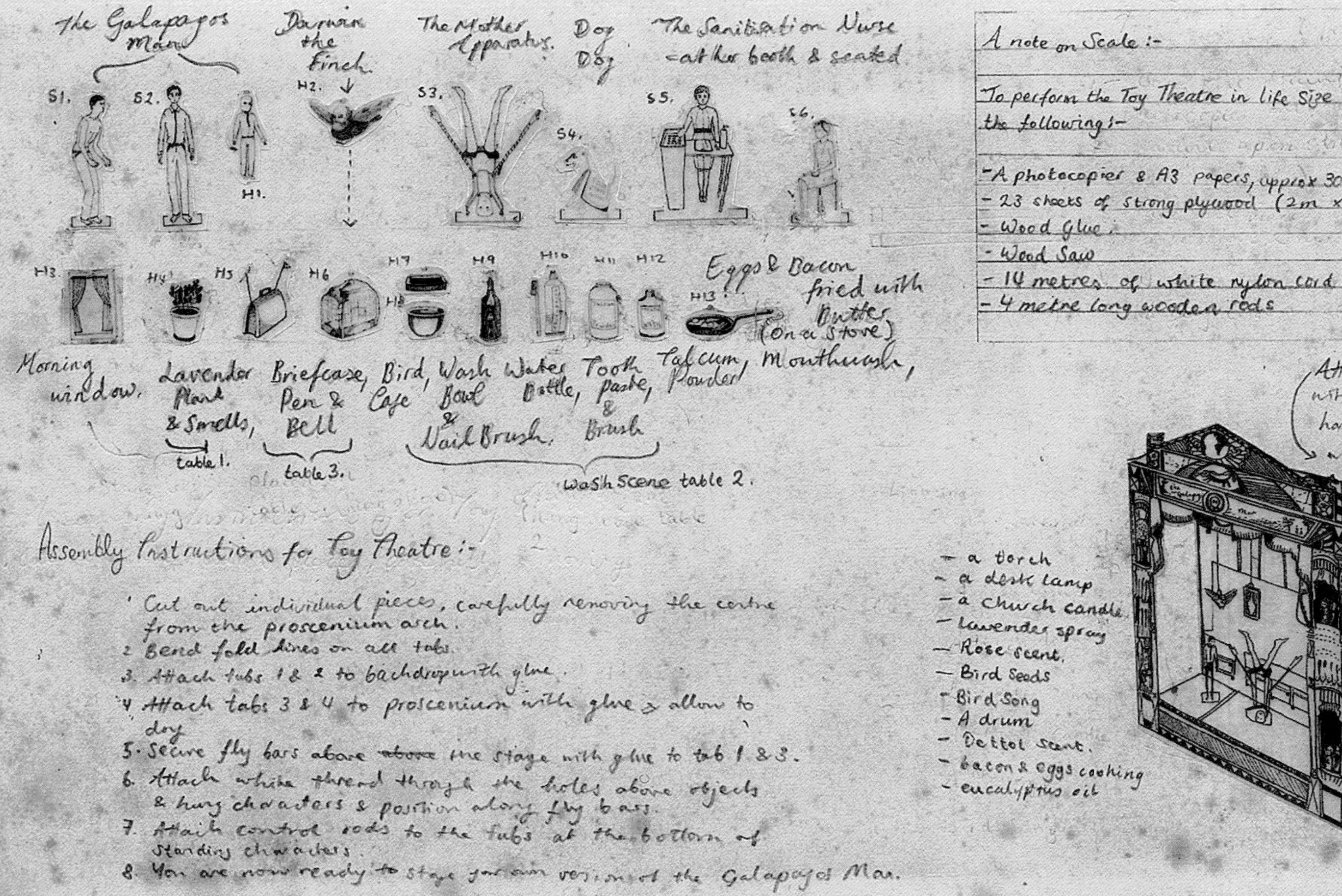

The Galapagos Man
Darwin the Finch
The Mother Apparatus.
Dog Dog
The Sanitisation Nurse - at her booth & seated
S1.
S2.
H1.
H2.
S3.
S4.
S5.
S6.
H3.
H4
H5
H6
H7
H8
H9
H10
H11
H12
H13
Eggs & Bacon fried with Butter (On a Stove)
Morning window.
Lavender Plant & Smells,
table 1.
Briefcase, Pen & Bell
table 3.
Bird Cage
Wash Bowl & Nail Brush.
Water Bottle,
Tooth paste, & Brush
Talcum Powder
Mouthwash,
Wash Scene table 2.
A note on Scale :-
To perform the Toy Theatre in life size you will need the following :-
- A photocopier & A3 papers, approx 300 sheets.
- 23 sheets of strong plywood (2m x 1m)
- Wood Glue.
- Wood Saw
- 14 metres of white nylon cord.
- 4 metre long wooden rods
Attach 2 fly bars with glue, for hanging objects with white cotton thread.
tab 3
tab 4
- a torch
- a desk lamp
- a Church candle
- Lavender spray
- Rose scent.
- Bird Seeds
- Bird Song
- A drum
- Dettol scent.
- bacon & eggs cooking
- eucalyptus oil
Assembly Instructions for Toy Theatre :-
1 Cut out individual pieces, carefully removing the centre from the proscenium arch.
2 Bend fold lines on all tabs.
3. Attach tabs 1 & 2 to backdrop with glue.
4. Attach tabs 3 & 4 to proscenium with glue & allow to dry
5. Secure fly bars above the stage with glue to tab 1 & 3.
6. Attach white thread through the holes above objects & hang characters & position along fly bars.
7. Attach control rods to the tabs at the bottom of standing characters.
8. You are now ready to stage your own version of the Galapagos Man.

W

Girls Interrupted
Gendred spectres. Atlantic drag

KATHLEEN M. GOUGH

> The spirit, the spectre are not the same thing and we will have to sharpen this difference, but as for what they have in common, one does not know what it is, what it is presently.... One does not know: not out of ignorance, but because this non-object, this non-present present, this being there of an absent or departed one no longer belongs to knowledge. At least no longer to that which one thinks one knows by the name of knowledge.
>
> Jacques Derrida (1994: 6)

> If she 'is' anything, she is the unconscious of the law, that which is presupposed by public reality but that cannot appear within its terms.
>
> Judith Butler (2000: 39)

The two epigraphs that preface this essay enact the vexed relationship between spectres and knowledge, and women and the law. In contemplating the relationship between the spirit and the spectre, Derrida reminds us that 'one does not know what it is' because this relationship, this 'non-present present ... no longer belongs to knowledge'. The anxiety surrounding what it 'is' - in the first instance - refers to the inability to distinguish between the spirit and the spectre. 'It' also seems to manifest a related referent: the absent presence of some unknowable 'other'. 'It' (and what it is) is unknown, in part, because it is not legible; it seems ungraspable and thus (one could assume) not particularly relevant as a subject of historical investigation.

In *Antigone's Claim*, an illuminating study of Antigone's multifarious philosophical, theoretical and cultural deployments, Butler contemplates the relationship between kinship and the state, and women and the law, by considering what Antigone is. In a register that echoes Derrida, Butler is also attempting to grasp at the ungraspable: 'if she "is" anything', we are told she is 'that which is presupposed by public reality but that cannot appear within its terms'. She is, in effect, a spectre. To invoke Derrida once again, she is part of the 'haunting [that] belongs to the structure of every hegemony' (1994: 37).

Thus, I begin this discussion about civil rights and haunting in the Atlantic world by suggesting that if Derrida's *Spectres of Marx* provides Derrida and us with some 'real' and 'theatricalized' ghosts that prove unwieldy and unworldly, Butler alerts us to the ways that female-gendered spectres compound this conundrum exponentially. Although Antigone is unable to appear within the terms of public reality, Butler explains how 'as a figure for politics, she points somewhere else, not to politics as a question of representation, but to that political possibility that emerges when the limits of representation and representability are exposed' (2001: 2).

It was precisely this consideration of the political possibility that emerges at the limits of representation that led me to take seriously the efficacy of gendered spectres. For quite some time now I have been researching the political and cultural activities among and between African-American and Irish Catholic communities during the American and Northern

Performance Research 13(4), pp.115-126 © Taylor & Francis Ltd 2008
DOI: 10.1080/13528160902875705

Irish civil rights movements. Journalistic and scholarly articles, photos, memoirs, plays, edited volumes and historical monographs all indicate explicitly or tacitly that African-American women in the Southern U.S. effectively started the civil rights movement, and it was the activities of these women that the first civil rights demonstrators in Northern Ireland - Irish Catholic women - looked to for civil rights models. Yet, these women on both sides of the Atlantic keep disappearing from critical view. I've been wondering what to make of this 'body snatching'.

Despite an abundance of scholarship on either side of the Atlantic (in local and national frames) where scholars have indeed performed the tasks of locating, retrieving, indexing, imaging and imagining women's central role in the movements (a sizeable amount of archival materials have been amassed), women still appear to disappear, or appear to be marginalized and interrupted, in grand narratives of the movements, in comparative scholarship of the movements, and as the central instigating forces behind and at the forefront of the movements. To read women's cultural and political performances *as* disappearance, *as* disappearing, however, are we 'limiting ourselves to an understanding of performance predetermined by our cultural habituation to the logic of the archive?' (Schneider 2001: 100). The ways that the archive has failed to acknowledge the power of the performing spectre as 'the unconscious of the law' is an issue to which I will return in several guises throughout this essay.

While my research began with an examination of political and theatrical performances of, by and about women in the civil rights movements, their alleged disappearance (or interruption) has far-reaching consequences for considering the absence of women in scholarship of the Black and Green (Irish) Atlantic more generally. Here I examine how performance helps to illuminate gendered blindspots in comparative studies of Atlantic cultural production.

One of the animating principles for theorizing this absence comes from a preoccupation with thinking through the political efficacy and pitfalls of what I call the 'liberation ontology' that frames many narratives of transnational politics and intercultural performance in the Black and Green Atlantic. Studies that explore the political affiliations and cultural relationships between African-Americans and Irish Catholics *typically* follow a diachronic path: Frederick Douglass's and Daniel O'Connell's anti-slavery and anti-colonial affiliations in the mid-nineteenth century have - in many respects - become the *ur*-text that then helps to account for the theatrical and literary relations in the early twentieth century (the Irish and Harlem Renaissances) and also the civil rights political affiliations in the mid-twentieth century.

While not denying the important work O'Connell and Douglass accomplished and the ways their twinned-legacy has been put to productive political and critical use, I do not want to place them at the forefront of a something akin to a liberation ontology. Instead I want to try to situate the problems of comparative civil rights historiography and women's position in this corpus of work in a differently inflected frame. This is one that Rebecca Schneider explains, in her essay 'Archival Remains', is not predicated on '*presence* that appears in performance but precisely the missed encounter - the reverberations of the overlooked, the missed, the repressed, the seemingly forgotten' (2001: 104).

I consider two spectres that ghost the Douglass-O'Connell alliance: Aunt Jemima and Cathleen Ní Houlihan. I am interested in what Aunt Jemima and Cathleen Ní Houlihan do to ontology and to the politics of this particular liberation ontology. If we are to understand - or at least entertain the idea - that Aunt Jemima and Cathleen Ní Houlihan, fictive icons made famous on stages in America and Ireland in the wake of American Slavery and the Great Irish Famine, persist in affecting how women are represented in comparative accounts of the Black and Green Atlantic, then I want to suggest that

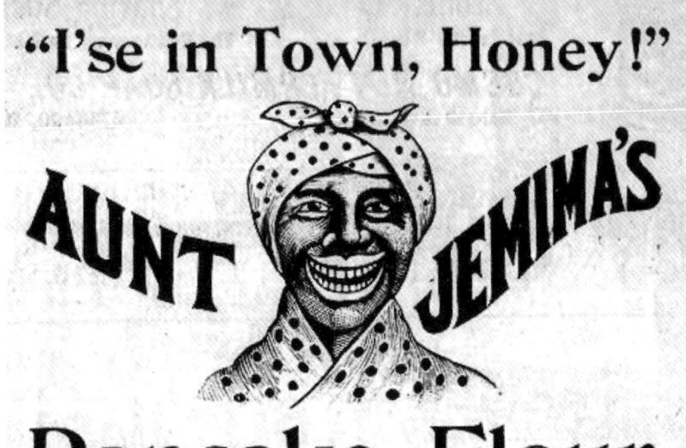

• Figure 1. This advertisement for Aunt Jemima's Pancake Flour appears in a programme for the Ed. F. Davis *Uncle Tom's Cabin* Company, which played at Cincinnati's Pike Opera House in early 1894, less than a year after 'real-life ex-slave' Nancy Green starred as Aunt Jemima at the Chicago World's Fair. The image of Aunt Jemima depicts the conflation of the blackface drag minstrelsy tradition alongside the appearance of the 'Original Aunt Jemima' baking pancakes at a local department store. Image published with permission of the owner, Mary C. Schlosser, and the 'Uncle Tom's Cabin and American Culture' Multi-Media Archive at the University of Virginia

these characters, turned icons, turned myths, turned spectres are not simply kitschy ephemera whose very existence is superfluous to historical inquiry. Instead, these spectres function as a kind of camouflage which, when taken seriously, can help to refigure logics of time and space: they produce insight not only in terms of *content*; these spectres can help to trouble the framework of nationalist historiographies more generally.

Before considering this idea in more detail, it might be useful to outline a brief genealogy of these two infamous 'women'. In *Slave in a Box: The Strange Career of Aunt Jemima*, M. M. Manring gives a very lucid account of her very strange career. In 1889,

> [s]he began as a white man, in drag, wearing a blackface, singing on a minstrel stage. She became a face on a bag of pancake flour, then a real-life ex-slave [Nancy Green] who worked in a Chicago kitchen but cobbled together enough reality and fancy of life under slavery to entertain crowds of the 1893 World's Fair. Next, advertising copywriters and one of America's most distinctive illustrators brought her to life in the pages of ladies' magazines, although the print version was still shadowed by a succession of real-life Aunt Jemimas who greeted the curious public at county fairs and club Bake-Offs. Then, in the era of the Civil Rights movement and Black Power, she grew into a liability, someone to be altered to meet the times and explained away with no small amount of embarrassment (1998: 1).

While Aunt Jemima's strange career is invoked by her many mammy manifestations, Cathleen Ní Houlihan served – and in some respects continues to serve – as the militant hag-maiden of the Irish Nationalist cause. Richard Kearney explains, as a Poor Old Woman she

> summons her faithful sons to rise up against the infidel invader so that, through the shedding of their blood, she might be redeemed from colonial violation and become pure once again – restored to her pristine sovereignty of land, language, and liturgy. (1997: 120)

Though the figure of Cathleen was first adopted by Jacobite poets in the eighteenth century, Cathleen's physical and symbolic call to the men of Ireland was made (in)famous by actress-revolutionary Maude Gonne who played Cathleen Ní Houlihan in W. B. Yeats's and Lady Augusta Gregory's 1902 play of that same name. She was the definitive icon of Ireland during the 1916 Easter Rising and in the subsequent Irish War for Independence (1919–21), and her battle cry resounded in the six counties of Northern Ireland throughout the mid- and late-twentieth-century Troubles.

One of the things that may seem askew – and it is necessarily askew – in taking Aunt Jemima and Cathleen Ní Houlihan as figures that have never really belonged to knowledge (or as Derrida reminds us, at least 'that which one thinks one knows by the name of knowledge' [1994: 6]) and suggesting that they can help us produce it, is that they seem so particular. Or at least they do not seem to circulate in the same representational economy. Indeed, whereas Aunt Jemima is reproduced as commodity – and commodity fetish – Cathleen Ní Houlihan reproduces allegory. Where Aunt Jemima works by way of ubiquitous reproduction, Cathleen Ní Houlihan works by way of discrete disappearances. In other words, when the call to Cathleen (nationalism) is heralded, 'real' women seem to vanish.

• Figure 2. A scene from W. B. Yeats's and Lady Augusta Gregory's *Cathleen Ní Houlihan*, circa 1912.

The gendered relationship between commodity and allegory in this instance (and one could argue that from Defoe's 'Lady Credit' onward the relationship between commodity and allegory was always gendered) has much to tell us about the relationship between reproduction, disappearance and historiographical practices in ways that trouble ontological thinking. That is, the moving threshold between commodity and allegory demonstrates how the thing that is reproduced as non-value - we could call this 'thing' Aunt Jemima or Cathleen - (allegedly) disappears because it is perceived as valueless.

In her discussion of Walter Benjamin's theory of allegory, Theresa Kelley states that 'emblem and commodity are, along with personification and stereotype, dialectical images whose abstractions or abstracting energies are lodged in material and visual forms that gesture beyond themselves, frequently into the past' (1997: 253). She goes on to suggest that whether 'allegory or commodity, the double signpost or dialectical image of this backward glance is ruin' (253). I suggest that both Aunt Jemima and Cathleen Ní Houlihan are these dialectical visual and material images whose 'abstracting energies' gesture to the past - be it to American slavery or to the Great Famine (1845-51) - that is figured as (gendered) 'ruin': that which is 'unspeakable' or 'unrecoverable' but nevertheless reproducible in reified form.

The very ubiquity of Aunt Jemima's reproduction (from minstrel stage to pancake flour bag to the 1893 World's Fair to magazine images to Club Bake-Offs to the 1960s policies on the black family led by Daniel Patrick Moynahan) means that black female subjectivity is disappeared under the weight of the spectre's prolific reproductions. Concomitantly, the very singularity of Cathleen (she appears under various guises such as Maude Gonne and later Northern Irish political activist Bernadette Devlin, but she is always a lone rider) means that Irish female subjectivity is reproduced as disappearance. Thus, Gonne or Devlin *appear* as the exception that allows for the category of 'woman' to be excluded. So one form of subjectivity is disappeared through the (commodity) spectre's reproductions, and another form of subjectivity is reproduced by way of and *as* (allegorical) disappearance.

One of the reasons this seems so complex is the double-duty of mimesis in producing political effect and theatrical affect. Let me explain why this might be problematic by referring to Elin Diamond's carefully theorized articulation of the dilemma. There is a long tradition 'that interprets mimesis as not only the act of imitating but an imitation *of*' (1997: iii), the embodied activity of representing and the representation. Responding to an earlier statement made by Richard McKeon (1952: 122), Diamond summarizes:

'Imitation indicates a constant relation between something which is and something made like it. ... ' Tangled in iconicity, then, in the visual resemblance between an originary model and its representation, mimesis patterns difference into sameness. (1997: iii)

Following this succinct explanation, Diamond asks the essential question: 'Same as what?' (1997: iv). This, of course, is the conundrum that lies at the centre of feminist critiques across the disciplines but is exacerbated in comparative studies where thinking about analogous relationships, similarities and 'sameness' is, most often, the modus operandi (i.e., one political activist's relation to another; one kind of nationalism being made analogous to another; even two disparate modes of feminist critique being fashioned into a single model). Thus, comparative studies enact and manifest the meta-crisis of mimesis' double-duty, making its proliferating registers all the more difficult to track.

Before getting ahead of myself, I first wish to consider some localized examples that help lay the groundwork for understanding the implications of this 'meta-crisis'. Consider Maude Gonne's performance of that militant hag, Cathleen Ní Houlihan, in light of Diamond's comments. It would seem that in the aesthetic or semiotic sense Cathleen Ní Houlihan would be the representation (Gonne providing the embodied activity of representing), but wouldn't Cathleen Ní Houlihan also be the originary model

• Figure 3. Maude Gonne, circa 1900.

(could she be both the origin and the representation or would this somehow make Gonne the representation)? The problem becomes who is imitating whom, or 'the same as what?'

So, what if the logic of this argument is faulty? After all, Maude Gonne was an actress playing a part; 'origins' are bound to be complicated. However, Gonne was also an Irish Revolutionary (in the political valence of representation she acted on behalf of other persons). 'In 1902 Maud Gonne was at the apogee of her career as an Irish nationalist' (Quinn 1997: 46). On stage, however, she could only play the figure of revolution (the figure that incites revolutionaries) and the violated land that necessitates the revolution in the first place. She cannot play the revolutionary herself. Yet, the fact that she *was* a revolutionary certainly added to the 'authenticity' of the Cathleen performance (and Cathleen's 'life' outside the theatre) in many complicated ways - in fact, in ways that leads one to consider that perhaps Maude Gonne was the originary model after all. But as she is in no way the violated land itself, this doesn't quite make sense.

What about examples from another register of social performance: the performances of countless female civil rights activists in Northern Ireland in the early 1960s and the political performances of Bernadette Devlin in the late 1960s and early 1970s. I will simply note here that these unreliable (or to use a phrase from Margaret Ward) 'unmanageable revolutionaries' never appeared where they should. They were on the street when they should have been in the home, or in the case of Bernadette Devlin, she was in America working with the Black Panther Party when she should have been in university singing political folksongs. Of course, the real problem is that there is no 'proper' place for activist women who work outside the confines of traditional constructions of nationalism to be fully represented in these conventional narratives. Or to put this another way: the 'proper' place - where women's political work 'counts' and their actions are 'visible' - would be 'the same as what?'

Their work is not always legible, and thus their unprecedented feats (i.e., ushering in the civil rights movement by constituting the domestic sphere as a political space in Dungannon in 1963 or travelling abroad to form anti-racist socialist alliances outside of the nationalist frame) are often intractable. Their promiscuous and threatening alliances and movements disrupted a certain (ethno-national) formal coherence and are reflected back as simply a contained and containable domestic disturbance - white noise to the real agents of national and trans-Atlantic social change.

The spectre of Cathleen Ní Houlihan has much to answer for in this regard. But herein lies the complication of mimesis and mimicry. To return to Elin Diamond's discussion of this dilemma, she tells us:

> Mimesis ... is impossibly double, simultaneously the stake and the shifting sands: order and potential disorder, reason and madness. As a concept mimesis is indeterminate and, by its own operations, loses its conceptual footing. On the one hand it speaks to our desire for universality, coherence, unity, tradition and on the other it unravels that unity through improvization, embodied rhythm, powerful instantiations of subjectivity and, what Plato most dreaded, impersonation, the latter involving outright mimicry. (1997: v)

There are many reasons to dread 'impersonation' and 'outright mimicry' in ways that are far removed from Plato's position. Thus, part of the story of women, haunting and civil rights in the Black and Green Atlantic (a narrative that cannot employ the metaphor of Gilroy's ship so much as something akin to a shipwreck) is that almost as soon as Cathleen's shadows walked out of the house and onto the streets (or flew over the Atlantic and landed in America), Cathleen's boys were there to ensure that the 'improvisation', 'embodied rhythm' and 'powerful instantiations of subjectivity', were reflected back as 'universality', 'coherence', 'unity' and 'tradition'. In this sense, Cathleen is the ultimate drag performer: fierce, powerful, inciting and exciting men to the cause, and disappearing any 'real' women who may get in her path (or step out of theirs). This is not, however, subversive drag but colonial mimicry: the double-edged sword, the other side of ambivalence articulated by Bhabha's differently inflected 'resemblance and menace' (1994: 86).

If Cathleen is the ultimate drag performer (after all, no one knows of her gender trouble) Aunt Jemima is - from her very inception - a failed drag act, but one that fails affectively. Like Cathleen Ní Houlihan's most popular manifestation in Yeats's and Gregory's 1902 play, Aunt Jemima began her life on the stage in 1889. More to the point, 'she began as a white man, in drag, wearing blackface, singing on a minstrel stage' (Manring 1998: 1). In a single embodied imitation of an imitation the white male performer manifested not only gender trouble, but race trouble, class trouble and kinship trouble. Indeed, a whole world of trouble. Blackface drag is a refracted form of mimesis: it is always hypervisible as impersonation; the blackface drag act fails so that the 'joke' can succeed. Yet something here still casts a mimetic shadow: Aunt Jemima would have to serve as the originary model for the representation, but she doesn't really exist. Like Cathleen, she began as a fiction with some fairly strange ramifications for 'real' life. Particularly considering that by 1893 Aunt Jemima was being played by real-life ex-slave Nancy Green, who worked in a Chicago kitchen and entertained crowds at the World's Fair - thus, making literal Fred Moten's redeployment of Marx when he discusses 'the commodity who speaks' (2003: 8).

In the remainder of this essay I want to return to the question of what Aunt Jemima and Cathleen Ní Houlihan do to the politics of the 'liberation ontology' that begins with Frederick Douglass and Daniel O'Connell. In turn, I wish to reflect on the chances and challenges for considering the appearance (and disappearance) of gendered performances in comparative Atlantic historiography. In this exploration I 'jump scales' to highlight the meta-crisis of mimesis' double-duty to explore how it operates in comparative studies. This examination circulates around questions of evidence, value, authenticity and 'the real'.

SLAVE WOMEN, FAMINE WOMEN AND EVIDENCE: TOWARDS A THEORY OF VALUE

> I see you now for what you are
> My wicked images. My simulacra.
> Anti-art: a foul skill
> traded by history
> to show a colony
> the way to make pain a souvenir.
>
> Eavan Boland, from 'Imago' (1998: 21)

Aunt Jemima and Cathleen Ní Houlihan were brought to life on the stage as 'innocent amusements' (*pace* Saidiya Hartman) in the wake of unfathomable historical violence - American Slavery and the Great Irish Famine. Violence that - at the scale of the body all the way to the scale of the colonial encounter - was always gendered female. Moreover, if these two figures were brought to life in the wake of historical violence, they were also brought to life in the wake of Douglass's and O'Connell's significant mid-nineteenth-century civil rights campaigns. That is, they appear after the political alliance and *after* the 'end' of slavery and the famine. In fact, because they appear after these historical moments, it would seem to imply that their subsequent appearance is anachronistic, inappropriate to consider in light of the Douglass-O'Connell 'liberation ontology'.

I suggest that the theory of knowledge and the form of value that made possible Aunt Jemima and Cathleen Ní Houlihan precedes and exceeds this ontology in ways that are still palpable in popular cultural practice and historiography. In taking a cue from Ian Baucom's illuminating discussion of the work of 'credit' in the early Atlantic world (see Baucom 2005: 14-17), my point is to note how much of the recalcitrant nationalisms sweeping through the Southern U.S. and Ireland in the late nineteenth and early twentieth centuries required both a theory of knowledge and a form of value that would secure the machinations of Southern (Confederate) nationalism and Irish nationalism to be in place. 'Central to that theory was a mutual and system-wide determination to credit the existence of imaginary things' (2005: 17). Two of the imaginary 'things' credited in this instance are, of course, Aunt Jemima and Cathleen Ní Houlihan. However, their imaginative and 'fictional' nature continues to hold material power with very real consequences. Indeed, I would argue that the power is all the more insidious for its seemingly evanescent existence.

What is valued - in both instances - is a bourgeois nationalism built upon the machinations of colonialism; the determination to give credit to the homogeneity of 'national' identity required the existence of an 'other' that was imaginary yet so constitutive to the building of a 'national' imagination that it could appear without comment. This gendered ('other') spectre is, as Butler reminds, 'that which is presupposed by public reality but that cannot appear within its terms' (2001: 39). Long before the visual and material appearance of Aunt Jemima and Cathleen Ní Houlihan, the theory of knowledge that made them possible was already secured, the value was already 'credited'. To understand the 'real' value of Aunt Jemima and Cathleen is to begin to recognize the 'real' values that underpin certain ideologies that continue to circulate - sometimes without intention - in critical methodologies out of which pan-nationalist movements on radically different scales and with different economic, political and cultural consequences are recorded, compared, written and theorized.

In order to consider this idea more fully I want to return to a discussion of the relationship between commodity and allegory, reproduction and disappearance. As Daphne Brooks points out,

> black women's bodies continue to bear the gross insult and burden of spectacular (representational) exploitation in transatlantic culture. Systematically overdetermined and mythically configured, the iconography of the black female body remains the

central *ur*-text of alienation in transatlantic culture. Hegemonic hermeneutics consistently render black women's bodies as 'infinitely deconstructable "othered" matter'. (2006: 7)

An exemplary reminder of the rendering of 'black women's bodies as "infinitely deconstructable 'othered' matter"' is no more clear than when we consider '[t]he fact that Aunt Jemima may have originally been "a white male businessman's appropriation of a 'German' male vaudevillian's imitation of a black male minstrel's parody of an imaginary black female slave cook"' (Morgan 1995: 88). Indeed, this attests to the character's origins as a commodity fetish. Aunt Jemima's uncanny ability to be a commodity fetish who could help sell pancake flour to Northerners by personifying 'southern hospitality' (1995: 88) during an era marked by the growing memory narrative of the 'Lost Cause', in which 'Southerners had a noble cause and had no reason to feel ashamed' (Bodnar 2000: 31), gives us some leverage for considering the theory of knowledge (economic, historic and national) that existed well before she was given life on the stage. Thus, black female subjectivity is disappeared under the weight of the commodity's ubiquitous reproductions.

In writing about allegory Walter Benjamin states that 'it is by virtue of a strange combination of nature and history that the allegorical mode of expression is born' (2003: 167). In Margaret Kelleher's groundbreaking study *Feminization of Famine*, she illuminates this 'strange combination' when she writes:

> The spectacle of famine, as early as the 1840s, is thus frequently constructed through female figures, its traces inscribed on hunger-ravaged, unclothed bodies. Women's function as 'bearer of meaning', in these famine texts, possesses a number of aspects. Most obviously, the maternal body displays the absence of nourishment, the failure of 'primal shelter' and sustenance ... The maternal body, with its particular associations with the 'unspoken' or 'non-speech', can thus function as the 'strange fold that changes culture into nature, the speaking into biology' [Kristeva 1985: 182]. The individual figure stands for a general breakdown or crisis, not only in society, but also in representation itself. (1997: 29)

Certainly, the Great Famine - as both a natural and political crisis - is one example of how Cathleen Ní Houlihan becomes the belated manifestation of how 'the allegorical mode of expression is born'. That is, the value that gave rise to Cathleen's performative emergence was evident before *what* was valued became fully apparent. As David Lloyd explains, the figure of 'Cathleen Ní Houlihan, often regarded by Irish revisionists as an atavistic female figure who mobilizes an irrational and regressive nationalism's devotion to oedipal emotions, is in fact a product of modernity and its insistence on homogenization of national identity' (1999: 36).

The long after-life of Aunt Jemima and Cathleen Ní Houlihan well beyond American slavery and the Great Famine allows us to consider how they are linked to a historical violence that (often unconsciously) but habitually structures the form that history - and historical writing - takes. In the absence of the voices and bodies of subaltern women, who for centuries circulated in the Black and Green Atlantic, Aunt Jemima and Cathleen Ní Houlihan appear as the camouflage blocking alternative pathways for conceptualizing time and space. That is, they become constitutive of a nationalist structure in which we cannot conceive of a life lived otherwise. Even as they haunt the ontology of so many circum-Atlantic narratives in and through the mid-twentieth-century civil rights struggles, they nevertheless materialize as painful souvenirs.

DOUGLASS, O'CONNELL AND THE 'REAL': TOWARDS AN UNDERSTANDING OF THE ARCHIVE

> People like they [sic] historical shit in a certain way. They like it to unfold the way they folded it up. Neatly like a book. Not raggedy and bloody and screaming.
>
> Suzan-Lori Parks (2001: 50).

The historical narrative - now well-rehearsed, but no less compelling - tells us in 1845, at the beginning of the Irish Famine, Frederick

Douglass escaped from Boston as a fugitive slave on a ship called The Cambria. Headed for Liverpool, Douglass stayed there for one night before embarking for Ireland where he spent nearly six months selling thousands of copies of his autobiography and lecturing against slavery to rapturous crowds of supporters. It was in Dublin that he also met the 'Irish Liberator' Daniel O'Connell, who introduced Douglass at a crowded Repeal meeting in September of that year as the 'Black O'Connell of the United States'.

As the famine in Ireland continued to get worse, however, O'Connell's concerns for American slavery and his somewhat dubious relationship with the Whig party meant that he was losing support among his followers in Ireland. Where the Young Irelanders were quick to condemn him, however, his support among abolitionists in America did not wane. Indeed, as Bruce Nelson points out, 'it was in America, not Ireland, that the radical O'Connell survived. For the Garrisonians, O'Connell was Ireland' (Nelson 2007: 81).

If O'Connell's radical anti-slavery stance survived in America, it can be argued that Douglass's radical internationalism - his cross-cultural and cross-political alliances - survive most visibly in Ireland. Indeed, Douglass's first visit to Ireland holds an important place in his development as a thinker and speaker. In the second edition of his autobiography, he writes that in Ireland he was 'not treated as a colour, but as a man - not as a thing, but as a child of the common Father of us all' (Douglass 1950: 120). Narratives of Douglass's visit to Ireland, alongside a play written in 2005 about his voyage on The Cambria and a recent mural of Frederick Douglass on the Solidarity Wall in Northern Ireland, all attest to his rather unprecedented comeback in both scholarly and popular media in Ireland over the last five years. In an era of mass globalization, increasing xenophobia and racism, Douglass, it seems, returns to the scene as a compelling and efficacious symbol of trans-Atlantic liberation.

I am not interested here in discounting the 'discrete' historical 'remains' that give rise to these accounts of Douglass's famous meeting with Daniel O'Connell in 1845, Douglass's anti-slavery lecture tours through Ireland or O'Connell's provocative political stance in mid-nineteenth-century America. I am more interested in reflecting briefly on the unprecedented return in comparative scholarship and popular representation to this historical moment - where Douglass and O'Connell become our primary witnesses to a time of complex solidarity, ambivalent compromise and trans-national liberation struggles. They, too, tell us something about the limits of representation. Paradoxically, their representational life circulates as 'real', 'authentic', somehow 'non-mimetic' (even when depicted in a mural or on the stage).

Take, for instance, the image depicted in figure 4. This mural of Frederick Douglass on the Solidarity Wall in Belfast portrays Douglass in the centre with a brief narrative of his alliance to Irish cause on the left and a quotation from his 1883 Civil Rights speech on the right. This speech was famous for - among other things - illuminating the exploitation of Ireland by the English before stating 'Fellow-citizens! We want no black Ireland in America' (Douglass 1883). Likewise, in Donal O'Kelly's 2005 play *The Cambria*, much of the text that Douglass speaks comes directly from his autobiography (and a substantial portion of these monologues are Douglass recounting his birth 'origins' and the beating of his Aunt Hester); the character 'Douglass', in effect, speaks his 'own words'. The

• **Figure 4. Mural of Frederick Douglass, Falls Road, Belfast, 2005.**

mural and the play represent Douglass as 'real' through the use of textual (archival) authority.

Indeed, the mural and the play re-enact the alliance and thus operate 'not [as] the historical acts themselves but the act of securing any incident backward' (Schneider 2001: 105). Schneider explains this process as a 'repeated act of securing memory' a means of rethinking 'the site of history in ritual repetition' (105).

The idea of securing memory and thinking of the site of history in 'ritual repetition' is, of course, central to the proliferation of studies, images and re-enactments of Douglass's and O'Connell's important political force in the Atlantic world. It is also a part of the ontological crises which Aunt Jemima and Cathleen Ní Houlihan manifest. It is both the capaciousness and pervasiveness of ontological thinking across time and space, however, that serves to keep one kind of representational economy in circulation while another remains either grounded or disappeared under the powerful forces of the Atlantic's drag. Indeed, it is the thrall to the concrete evidence, the strong intentionality of political influence, the physical, material, literal Atlantic crossings that the archive - in this instance and others - is so good at recording and that demands that other kinds of performances (like those of Aunt Jemima and Cathleen Ní Houlihan) disappear and continue to reproduce disappearance.

Yet, if we wish to 'rethink the site of history in ritual repetition' in order to explore the work that these knowledgeable spectres (often regarded as kitsch) do to trouble ontological and archival thinking, it is important to attempt a different understanding of saving, a method that takes seriously the reality of ideas and does not simply (or only) ascribe reality 'to what can be perceived directly, rather than experienced so to speak indirectly, viz., through its actual or potential effects, i.e., to a perceptual rather than causal criterion' (Bhaskar 1997: 139). In this way it might be possible 'to make room for the spectres in whose restlessness the rhythms of another mode of living speak to us' (Lloyd 2005: 153).

'Kinship trouble': thoughts on performance and 'authenticity' in the Black and Green Atlantic

In the final section of this essay I wish to offer some comments on the possibilities of studying the Black and Green Atlantic together. This will require a radical rethinking of how the notion of kinship may be used as a conceptual category to resituate comparative scholarship in an Atlantic frame. I want to begin with several questions. Would it be possible and productive to think of Aunt Jemima and Cathleen Ní Houlihan as kin? What purpose would it serve? What kinds of inroads could be made in thinking history and performance studies together? What kinds of ontological and hermeneutic structures would this alliance refigure?

Regarding these two figures as kin would allow us to approach them as 'authentic' visual and acoustic markers, which have produced 'real' effects and consequences. While they continue to operate as names and screens, we can utilize their efficacy for different ends: they, in fact, point towards the conditions of possibility for a situated examination of intercultural and inter-political entanglements in the Black and Green Atlantic.

I want to explain this by way of two brief examples. At an Irish Studies seminar in July of 2006, Kevin Whelan - Irish geographer, historian and cultural critic - began a discussion on the scholarship of Ireland in the greater Atlantic world by stating that Irish historiography has an 'anorexic relationship to its maritime conditions'. There is, indeed, *much* more scholarship on Ireland and nationalism than on Ireland in the Atlantic world. His reflections and his phrasing - however unintentional - cannot help but evoke the Great Famine (and the 'feminization' of that famine), 'American wakes', 'coffin ships', disappearance, an inability to circulate and an inability for the land, the language, the culture, the body to reproduce under such conditions.

When Irish historiography is figured as having an 'anorexic relationship' to the Atlantic, it is because that nineteenth-century rupture can never be fully accounted for, it closed in on itself. It left disappearance. A radical - and rabid - nationalism (to which Cathleen Ní Houlihan is constitutive) would follow in its wake. Thus, the condition of possibility for a Green Atlantic is disappearance (which nationalism continually reproduces).

• **Figure 5. Mural of Young Mother Ireland, Beechmount Avenue and Falls Road, Belfast, 1983.** *Image is published with permission of the artist, Bill Rolston.*

In Fred Moten's study *In the Break: The Aesthetics of a Black Radical Tradition*, he suggests that the condition of possibility for Black performance (which is not distinguishable from Black history in his argument) is reproduction. Indeed, he attempts 'to describe the material reproductivity of black performance and to claim for this reproductivity the status of an ontological condition' (2003:18). This, he tells us, 'is the story of how apparent nonvalue functions as the creator of value; it is also the story of how value animates what appears as nonvalue' (18). Moten instructively deploys the figure of Frederick Douglass's Aunt Hester (and her beating in chapter one of Douglass's autobiography) to discuss the 'the resistance of the object' and to elucidate his reflections on 'wounded kinship' - not as a suturing of but a resistance to claims of (biological) origins and (historical) authenticity.

Through this essay I have been arguing that we take seriously - and put more pressure on - the relationship between political effect and theatrical affect. The way these two notions inform each other continues to have far ranging consequences for how gender is made to 'count' and is valued. This is not so that we simply understand that politics informs theatre (the point we know) but that theatrical affect and the way it is deployed across a wide range of popular mediums and academic disciplines has a particular role to play in helping to illuminate gendered blindspots that continue to be elided in comparative studies. The rigorous theorizing in regards to mimesis that is often 'housed' in theatre and performance studies allows us to 'scale up' in order to consider the consequence of methods (always political - even if unintentional) that continue to value sameness and concrete 'evidence'.

How does the relationship between reproduction and disappearance, manifested in Aunt Jemima and Cathleen Ní Houlihan, become part of the story of nonvalued objects who create hegemonic value and of the story of the contradictory ways that this same 'value' animates these figures? Like Antigone, these two figures appear to occupy a space 'anterior to the state and anterior to kinship' (Butler 2000: 31). Their space is that of the Atlantic world's manifold contradictions. They haunt history as the Hegelian unconscious, they are resistant to dialecticism. They both echo and anticipate a number of other gendered spectres in the Atlantic unconscious. If Aunt Jemima is 'kin' with several famous figures also given life on American stages and in American fiction, Topsy and Dilsey (see Johnson 1998: 36), Cathleen Ní Houlihan is their shadow: in each case, they never 'was' born, they just endured.

Earlier versions of this essay were delivered at the Theatre Research Seminar series in the Department of Theatre, Film and Television Studies at the University of Glasgow (2007); The Keough Institute for Irish Studies Irish Seminar, O'Connell House, Dublin (2007); and the PSi Conference at New York University (2007). I am grateful for the feedback I received from the participants at these events. I would also like to express my thanks to Richard Gough as well as to Karen Lury, Catriona Macdonald and P. A. Skantze who read this essay in its various draft stages and provided invaluable comments.

REFERENCES

Baucom, Ian (2005) *Specters of the Atlantic: Finance Capital, Slavery and the Philosophy of History*, Durham: Duke University Press.

Benjamin, Walter (2003 [1963]) *The Origin of German Tragic Drama*, trans. John Osborne, London: Verso.

Bodnar, John (2000) *Remaking America: Public Memory, Commemoration and Patriotism in the Twentieth Century*, Princeton: Princeton University Press.

Bhabha, Homi K. (1994) 'Of Mimicry and Man: The Ambivalence of Colonial Discourse', *The Location of Culture*, New York: Routledge, pp. 85-92.

Bhaskar, Roy (1997) 'On the Ontological Status of Ideas', *Journal for the Theory of Social Behaviour* 27.2/3: 139-47.

Boland, Eavan (1998) *Lost Land*, New York: W.W. Norton.

Brooks, Daphne A. (2006) *Bodies in Dissent: Spectacular Performances of Race and Freedom, 1850-1910*, Durham: Duke University Press.

Butler, Judith (2000) *Antigone's Claim*, New York: Columbia University Press.

Cowan, Bainard (1981) 'Walter Benjamin's Theory of Allegory', *New German Critique* 22 (Winter): 109-22.

Derrida, Jaques (1994) *Specters of Marx: The State of the Debt, the Work of Mourning and the New Internationalism*, trans. Peggy Kamuf, New York: Routledge.

Diamond, Elin (1997) *Unmaking Mimesis*, New York: Routledge.

Douglass, Frederick (1883) 'The Civil Rights Case, 22 October 1883' <http://teachingamericanhistory.org/library/index.asp?document=774>.

Douglass, Frederick (1950) *The Life and Writings of Frederick Douglass*, vol. 1, ed. Philip S. Foner, New York: International Publications.

Ferreira, Patricia (2001) 'Frederick Douglass in Ireland: The Dublin Edition of His *Narrative*', *New Hibernia Review* 5.1 (Spring): 53-67.

Johnson, Barbara (1998) *The Feminist Difference: Literature, Psychoanalysis, Race and Gender*, Cambridge, Massachusetts: Harvard University Press.

Kearney, Richard (1997) *Postnationalist Ireland: Politics, Culture, Philosophy*, New York: Routledge.

Kelleher, Margaret (1997) *Feminization of Famine: Expressions of the Inexpressible?*, Durham: Duke University Press.

Kelley, Theresa M. (1997) *Reinventing Allegory*, Cambridge: Cambridge University Press.

Kristeva, Julia (1985 [1976]) 'Stabat Mater' in Marina Warner *Alone of All Her Sex: The Myth and Cult of the Virgin Mary*, London: Vintage Books, pp. 182.

Lloyd, David (1999) *Ireland After History*, Cork: Cork University Press and Field Day.

Lloyd, David (2005) 'The Indigent Sublime: Spectres of Irish Hunger', *Representations* (special issue on Redress) 92 (Fall): 152-85.

Manring, M. M. (1998) *Slave in a Box: The Strange Career of Aunt Jemima*, Charlottesville: University of Virginia Press.

Morgan, Jo-Ann (1995) 'Mammy the Huckster: Selling the Old South for a New Century', *American Art* 9.1 (Spring): 86-109.

Moten, Fred (2003) *In the Break: The Aesthetics of a Black Radical Tradition*, Minneapolis: University of Minnesota Press.

Nelson, Bruce (2007) '"Come out of Such Land, You Irishmen": Daniel O'Connell, American Slavery and the Making of the "Irish Race"' *Eire-Ireland* 42.1/2 (Spring/Summer): 58-81.

O'Kelly, Donal (2005) *The Cambria: Frederick Douglass's Voyage to Ireland 1845*, <www.irishplayography.com>.

Parks, Suzan-Lori (2001) *Topdog/Underdog*, New York: Theatre Communications Group.

Quinn, Antoinette (1997) 'Cathleen Ní Houlihan Writes Back: Maud Gonne and Irish

Nationalist Theatre', in Anthony Bradley and Maryann Gialenella (eds) Gender and *Sexuality in Modern Ireland*, Amherst: University of Massachusetts Press, pp. 39-59.

Schneider, Rebecca (2001) 'Performance Remains', *Performance Research* 6.2: 100-108.

Ward, Margaret (1983) *Unmanageable Revolutionaries: Women and Irish Nationalism*, London: Pluto Press.

The Picturesque World Stage

GLEN MCGILLIVRAY

In 1781 the painter and stage designer Philippe Jacques de Loutherbourg presented a spectacle at his home in Lisle Street off Leicester Square. The *Eidophusikon* (Figure 1) was a box ten feet wide by six feet high and eight feet deep mounted in the wall of a room. This small theatre allowed Loutherbourg to concentrate light and combine the technologies of the magic lantern and shadow puppet shows with mechanised automata so that, when the room lights were darkened, a mini spectacle appeared before his audience's eyes. In miniature scale, Loutherbourg achieved what he had been unable to perfect over the last decade working on the London stage and that was a moving pictorial representation of the world in three dimensions (albeit somewhat reduced in size).

The series of vignettes displayed in this first season of the *Eidophusikon* consisted of scenic views at different times of the day and under moonlight, 'exotic' locales (Naples and Tangiers), and culminated with a stormy shipwreck. In a

• Figure 1: The Eidophusikon Showing Satan arraying his Troups on the Banks of a Fiery Lake with the Raising of Pandemonium from Milton. (Edward Francis Burney, 1782) © *The Trustees of the British Museum*

Performance Research 13(4), pp.127-139 © Taylor & Francis Ltd 2008
DOI: 10.1080/13528160902875713

second season in 1782 the vignettes comprised the sun rising through fog; Niagara Falls; the popular shipwreck again; a scenic view of Dover castle, the town and cliffs; a moonlit waterspout off a Japanese sea coast and culminated in 'SATAN arraying his TROOPS on the BANKS of the FIERY LAKE, with the Raising of the PALACE of PANDEMONIUM, from Milton' (Baugh 2007: 50). These 'scenographic tropes' would stimulate and inspire similar scenic inventiveness for the next half century (ibid.)[1]

[1] The Eidophusikon had three seasons: the first season opened on 26 February 1781 and the second on 31 January 1782, both in Loutherbourg's home. The third season, testifying to its popularity, was in the 200 seat Exhibition Rooms above Exeter 'Change and opened on 31 January 1786 (Baugh 1990: 103-104).

Loutherbourg's *Eidophusikon* emblematised a new aesthetic sensibility that had emerged during the course of the eighteenth century. English aristocrats travelling to Italy via the Swiss Alps on their Grand Tours, after avoiding hostile landscapes that threatened ambush by brigands or a sudden death plunge into the abyss, discovered dramatic landscapes and luminous light in paintings by the Baroque artists Salvator Rosa and Claude Lorrain. A sense of place flowed into them through their eyes and affected them emotionally; this was new: seeing could also be feeling. The world represented as landscape produced strong emotions in the 'sensitive' observer who, in turn, viewed actual scenery through the conventions of landscape painting. By the late eighteenth century, these conventions had become formalised and were known by a widely used and contested term: 'the picturesque'. After the publication of the Reverend William Gilpin's *Observations on the River Wye* (1782) picturesque tourism became the rage (Wales, the Lake District, and Scottish highlands being popular destinations) and 'the picturesque' became the key term - used and argued over by Humphrey Repton, Uvedale Price and Richard Payne Knight - in English landscape and garden design in the 1790s. Loutherbourg, as a pre-eminent artist of the picturesque, reflected this sensibility in both the subject matter and content of the scenes in his *Eidophusikon.*

In this essay I am interested in exploring the idea of theatre embedded within late eighteenth-century picturesque sensibility. If, as David Marshall writes, 'the picturesque represents a point of view that frames the world and turns nature into a series of living tableaux' then we are considering a mode in which the world appears to the spectator as distinctly theatrical (Marshall 2005: 17). But how and why did the language of theatricality take hold of pictorial sensibility when, for most of the century, the eighteenth-century stage created only a partial pictorial illusion? In addressing this question I argue that theatricality as an ontology, expressed through the Renaissance trope of *theatrum mundi*, rather than theatre as a practice, continued to inform eighteenth-century social and aesthetic sensibility (although it is only the latter of these that I will address in these pages). Despite losing its emblematic and allegorical resonances, *theatrum mundi* persisted into the eighteenth century and as it became secularised so too did it become pictorialised. The world was 'put in the picture' but that picture frequently organised the elements within its frame as though part of a theatrical scene and aesthetic theorists in England and France continued to use theatrical tropes in order to account for what they were seeing.

FROM THEATRUM MUNDI TO WORLD PICTURE

Discussions of theatricality throughout the eighteenth century in England and France return to the relationship between the truth of a thing and its outward appearance; in particular, the question of how a thing can be represented in its true state. Rather than simply reiterating Platonic arguments against mimesis, philosophers like the Earl of Shaftesbury and Denis Diderot instead wrestled with the idea of theatricalised existence and how, within it, one was able to truly represent either one's self or the world. David Marshall writes that for Shaftesbury, 'theatricality - the intolerable position of appearing as a spectacle before spectators - calls for the instatement of theatre: the protective play of masks and screens that would deny the view of the spectators it positions and poses for' (Marshall 1986: 67). Diderot, seeking a direct communication between his 'soul' and a painting, demanded a dramatic tableau the emotional intensity of which would arrest the spectator's attention; at the same time, through the absorption of its figures, the painting would seem to deny the spectator's presence in order to, paradoxically, invite his or her beholding (Fried 1980). Shaftesbury displays a similar paradoxical attitude to theatricality and theatre when, on the one hand, he claims that 'the good painter must ... take care that his Action be not theatrical, or at second hand; but original and drawn from Nature her-self', and on the other employs theatrical metaphors to extol nature that 'affords an ampler scene, and is a nobler spectacle than all which art ever presented' (quoted in Balme 2006: 3). Shaftesbury's unselfconscious concatenation of nature with theatrical spectatorship expressed a particular aesthetic sensibility that equated landscape with a theatrical scene while elsewhere, with typical suspicion of artifice, he condemned theatricality in artistic representation.

This theatricalised appearance of the world that delighted Shaftesbury (but which he objected to in painting) foreshadowed a new sensibility emerging from English and Continental aesthetic theory; a sensibility that endowed the natural world, through landscape, with new powers to affect the human observer. It is the spectator's relationship to the scene of nature that will exercise thinkers in the eighteenth century; in particular, its power to emotionally affect a spectator. Not only does nature acquire an affective power unknown to previous generations, but the spectator develops a heightened sensitivity to nature and the 'habit', in Christopher Hussey's evocative phrase, 'of feeling through the eyes' (Hussey 1967: 4). When the playwright and co-founder of *The Spectator*, Richard Steele, a contemporary of Shaftesbury, wrote in 1720 that 'the World and the Stage which have been ten thousand times observed to be the Pictures of one another', not only did he reiterate a hackneyed cliché but he reflected a growing sense that the world could, quite literally, be pictured (Steele 1720: 51)[2]. And, away from arts and letters, in the field of natural history, it was.

Developments in microscopy during the eighteenth century fulfilled, according to Barbara Stafford, 'the long-standing human yearning for visually entering entirely different realms' (Stafford 1993: 97). When the solar microscope, a device similar in operation to a magic lantern, was developed in the mid-eighteenth century it allowed the projection of magnified images onto a screen; no longer was microscopy a past-time for the solitary viewer, but could be shared with an audience (Stafford 1993: 98). It was not sufficient to simply report on the 'minute world' witnessed in the microscope's lens, but natural historians frequently orchestrated the 'sensational re-creation of the actual experience of witnessing' these miniature discoveries (Stafford 1993: 106). The natural historian, Henry Baker, wrote in 1744 of the dissected frog beneath his microscope: 'No Words can describe the wonderful Scene that was presented before our eyes;' moreover, he continued, the magnified circulation of its blood 'appeared like a beauteous landscape, where Rivers, Streams, and Rills of running Water are everywhere dispersed' (quoted in Stafford 1993:108,109)[3]. According to Stafford the:

> transubstantiation, by metaphorization, of bodily suffering into an inanimate landscape painting was a technique for distancing the beholder from the awareness of *how* what he beheld had been contrived ... This picturesque *tableau vivant* was literally suspended, or hung for viewing, in the gallery of the laboratory. (Stafford 1993: 109)

Distanced from the anguish of the suffering creature, the world that appeared to Baker and his enthralled audience was both theatrical scene and picturesque landscape.

Shaftesbury's 'ampler scene' and Baker's 'wonderful Scene' suggest a subtle shift in the Renaissance theatrical metaphor that emphasised its visual dimension. In both these

[2] According to Christopher Hussey, an early use of the term 'picturesque' appeared in Steele's play *The Tender Husband* (1723); however, it was not yet associated with landscape (1967: 32).

[3] From Baker, Henry (1744) *The Microscope Made Easy; or, The Nature, Uses, and Magnifying Powers of the Best Kinds of Microscopes Described, Calculated and Explained: For the Instruction of Such, Particularly, as Desire to Search into the Wonders of the Minute Creation, tho' They Are Not Acquainted with Optics*, 3rd edn, 2 vols, London: Dodsley.

examples the human actor is absent from the scene but has become the spectator looking on to the spectacle of nature. The origin of the theatrical metaphor that emerged in the 1500s can be found in the twelfth-century scholar John of Salisbury's *Policraticus* which was extensively re-printed during the sixteenth century.[4] *Theatrum mundi*, as it appears in the *Policraticus*, emphasises that it is the human being who is actor on this world stage and this performance is watched by an audience of sages, angels and God:

4 Ernst Curtius argues that the *Policraticus* achieved wide circulation in the Middle Ages but was also much read during the Renaissance, being reprinted in 1476, 1513 (in Paris and in Lyon), 1595, 1622, 1639, 1664, 1677 (Curtius1953: 140).

> They view the world-comedy along with Him who towers above to watch ceaselessly over men, their deeds and their aspirations; for since all are playing parts, there must be some spectators. [Therefore] Let no one complain his acting is marked by none, for he is acting in sight of God, of his angels, and of a few sages who are themselves also spectators at these Circensian Games (Pike 1938: 180).

There 'must be some spectators', writes John, and the Renaissance human is reassured (and warned) that all 'his' deeds are 'marked'. These divine spectators give the Renaissance theatrical metaphor its moral force; they are the ones who assess, weigh-up and judge, and before whom the human player will ultimately rise or fall.

From Dante's curious wanderer in the mid-fifteenth century through to Galileo's distanced observer in the early seventeenth, the relationship between human subject and the world shifted from immediate experience to abstracted contemplation. This transition marked the birth of the world picture which, as Martin Heidegger writes, did 'not mean a picture of the world but the world conceived and grasped as a picture' (Heidegger 1977: 129). No longer was the human subject a simple player in some divine comedy but had become a witness to the world also.

WILLIAM GILPIN AND THE 'PICTURESQUE EYE'

This theological dimension was not entirely lost by the eighteenth century - after all, the spectacle of the world was still God's creation - but the human being was now, even more, the spectator for whom the world puts on a show and during the eighteenth century the world would be more explicitly pictured than before. In an etching by Thomas Rowland, dated 1812, an elderly man

• **Figure 2: Dr Syntax Sketching the Lake (Thomas Rowlandson, 1812)** *©VandA Images/ Victoria and Albert Museum, London*

under a parasol, mounted on a grazing horse, sketches a lake while astonished rowers and a fisherman look on (Figure 2). The man, indicated by the cartoon's title, is Dr Syntax, a satirical rendering of the eighteenth-century exponent of picturesque art and travel: the Reverend Dr William Gilpin. In 1782, Gilpin published *Observations on the River Wye, and several parts of South Wales, & c. relative chiefly to picturesque beauty; made in the summer of the year 1770*, the first of eight works dealing with the picturesque. Gilpin wrote:

> The following little work proposes a new object of pursuit; that of not barely examining the face of a country; but of examining it by the rules of picturesque beauty: that of not merely describing; but of adapting the description of natural scenery to the principles of artificial landscape; and of opening the sources of those pleasures, which are derived from the comparison
> (Gilpin 1973 [1782]: 1–2).

Gilpin's project, as the above suggests, involved a new way of seeing nature; a sensibility that applied aesthetic principles to describing landscape. By describing what he saw according to 'the rules of picturesque beauty' Gilpin imposed upon the landscape the requirement for it to appear 'like a painting'. The necessity and desire to do so arose because, wrote Gilpin, 'the immensity of nature is beyond human comprehension. She works on a *vast scale*'; therefore, the artist's role was to spatially confine this immensity by adapting 'such diminutive parts of nature's surface to his own eye, as comes within its scope' (Gilpin 1973 [1782]: 18; emphasis in original). Christopher Hussey writes that Gilpin was not averse to suggesting 'improvements' to a scene where nature – whose 'genius' lay more in colour and tone than in composition – had fallen short. The artist could move or alter trees, replace spreading oaks with withered stumps or reshape a mountain into a peak; and although he could not add a castle, ruin or river where none existed, he could add a rustic cottage or alter the line of a river or road (Hussey 1967: 113,115). Gilpin, writes Hussey, provided a 'comical vision' by 'first abasing himself before nature as the source of all beauty and emotion; then getting up and giving her a lesson in deportment' (Hussey 1967: 114).

According to David Marshall, the term 'scenery' first appeared in print in relationship to landscape in William Cowper's poetry volume entitled 'The Task' in 1784 (Marshall 2005: 203 N.3). Gilpin had published his first essay on the picturesque only two years earlier but the idea of viewing landscape as a composed scene had been circulating for some years, at least since the middle of the century. Gilpin, over the next twenty years, formulated rules for picturesque composition that specified not only what content was deemed picturesque, but also how that content should be arranged in the frame so as to provide a receding perspective of foreground, mid-ground and background framed by 'side-screens':

> Every view on a river, thus circumstanced, is composed of four grand parts; the *area* which is the river itself; the *two side-screens*, which are the opposite banks, and mark the perspective; and the *front screen*, which points out the winding of the river (Gilpin 1973 [1782]: 8; emphasis in original)

The side-screens are a direct reference to staging conventions (more of which I discuss below) and it seems that Gilpin turns to theatre in order to create a sense of aesthetic distance in his picturesque view. In so doing, he avoids the vertiginous fall into the immensity of the sublime by limiting his view because, according to Walter Hipple, 'the picturesque ... since it so depends on the character of boundaries, can never be infinite' (Hipple 1957: 211).

Although Gilpin calls a stage into being through his compositional principles, it is in the step he takes before this, at the actual place he chooses to sketch or paint and make his tableau, that he theatrically 'frames the world'. It is as though he carries a scenic backdrop in his mind that influences how he will decide which view from nature appears picturesque and should be sketched and noted in his journal. Gilpin's journals provided the foundation for other travellers who, following in his footsteps and stopping to gaze at the same vistas, learned to view them in a similar way. Christopher Baugh writes that Loutherbourg toured the Lake District several times and, from the series of paintings he produced from a trip taken in 1783, it is highly probable that his itinerary was inspired by the route taken by Gilpin (Baugh 1990: 44-45). By the end of the century, picturesque tourism was well established in England with the Lake District being a particular favourite. Thomas West had published in 1778 *A guide to the Lakes: dedicated to the lovers of landscape studies, and to all who have visited, or intend to visit the lakes in Cumberland, Westmoreland and Lancashire*, a work that eventually ran to eleven editions over the next forty-five years. West's guide 'collected and laid before [the traveller], all the select stations, and points of view noticed by those who have made the tour of the lakes' (2-3, quoted in Bicknell 1981: 48).

The tour of the Lakes, was like a promenade performance created by nature and directed by Gilpin and other aficionados of the picturesque who not only told the traveller where to look but also how to, as Gilpin instructed: 'The first source of amusement to the picturesque traveller, is the *pursuit* of his object - the expectation of new scenes continually opening and airing to his view' (Gilpin 1972 [1794]: 47; emphasis in original). And, to assist and encourage the correct view, Gilpin and others frequently resorted to a Claude glass (Figure 3). About four inches in diameter and oval shaped, the Claude glass was a convex tinted mirror that gathered the reflected landscape into it. The dark glass muted the colours in the reflected image accenting its tonal value so that the resulting picture took on the light qualities of a painting by Claude Lorrain (Hussey 1967: 107).[5] To use it, the viewer stood with his or her back to the scene and gazed upon the reflected and tightly framed image. These images, Gilpin writes, reflected the landscape 'like the vision of the imagination; or the brilliant landscapes of a dream'; yet, the Claude glass lent them, also, 'a flatness, something like the scenes of a playhouse, retiring behind each other' (Gilpin 1973 [1791]: 225, 227). With the assistance of devices such as the Claude glass, the viewer developed what Gilpin termed the 'picturesque eye' (Gilpin 1972 [1794]).[6]

• **Figure 3: Claude glass, England, 18th Century,** *© VandA Images/Victoria and Albert Museum, London*

THE WORLD STAGE

The 'picturesque eye' could just as easily be termed the 'theatrical eye'. When Gilpin outlined his principles of painterly composition so that the spectator could organise his or her view of nature accordingly, these principles were based on a particularly theatrical way of seeing. It was not just how he used the terminology of stage design in his descriptions, but how the analogy of world and theatre - expressed through the idea of a 'scene' - was somehow ontologically appropriate. Although versions of the theatrical metaphor dated back to ancient times, its Renaissance manifestation had, by Gilpin's time, been circulating for two hundred years. To think about the world, then, was to think about it in theatrical terms: places became scenes within which characters - the rake, the bawdy wife, the ne'er-do-well, the bumpkin - performed their roles. Although still carrying the pejorative aspects of 'second handedness' in some discourses (exemplified by Shaftesbury, quoted above), this theatrical perception of the world in the eighteenth century, as Christopher Balme

5 Claude Lorrain, Salvator Rosa and Gaspard Dughet (called Poussin) were the seventeenth century landscape artists who inspired the picturesque. Claude sought the ideal form within closely observed details of nature and then represented this in his paintings of the Compagna and St Alban hills. In time, people began to seek landscapes that actually resembled his paintings (Hussey 1967: 10).

6 The other important viewing apparatus used extensively in the eighteenth century was the camera obscura. In addition to the common room-sized devices there were also a variety of portable versions. One of these that was bound into a book and called *Théâtre de l'universe* was created in 1750 (Stafford et al. 2001: 307-14).

argues, 'link[ed] the aesthetic and the moral into a wider concept' (Balme 2006: 14).

A theatrical representation of life did not necessarily produce a juxtaposition that highlighted some superior category of the real but, rather, expressed a heightened awareness of the social roles people played.[7] John Dixon Hunt observes that Hogarth 'moved in his paintings from actual theater scenes of Gay's *The Beggar's Opera* to subjects where private theatricals alert one to the ambiguous status of actor and audience and finally to the theater of life itself' (Hunt 1992: 73). Nicolas Poussin, a seventeenth-century landscape painter, the brother-in-law of Gaspar Dughet (known to the English as 'Poussin') who inspired the English picturesque taste, actually used a model theatre to organise the *mise-en-scénes* of his paintings (Bicknell 1981: 6; Hunt 1992: 114-115). Similarly, Salvator Rosa, produced a series of *capricci* whose figures, Erika Esau writes, 'clothed in armour and exotic robes, and characterized by extravagant gestures, are removed from the world of everyday experience' (Esau 1991: 42); or, in other words, are theatrical representations. Figures from the commedia dell 'arte were also favourite subjects and Rosa, according to Esau, even joined commedia troupes and performed the role of the buffoon, Corviello (Esau 1991: 57, n.17).

Yet the connection of landscape painting to the theatre of the world was made even more explicit through the latter's link to landscape gardening. The seventeenth-century Italian garden was the model for French and English gardens, writes Hunt, and in it, 'the visitor was no longer a passive spectator' but was led, as Henry Wootton declared 'by several *mountings* and *valings*, to various *entertainements* [sic] of his *sent* [sic], and *sight*' (quoted in Hunt 1992: 54; emphasis in original). Italian Renaissance garden architecture utilised ideas drawn from antiquity (often misunderstood) such as amphitheatres actually built into the garden, alcoves and niches, curved pavilions, statuary and fountains. In describing the gardens at Vauxhall and Ranelagh, Hunt writes that:

> the gardens were provided with many other items quite specifically associated with the theatre: there were the real vistas down the walks, assisted on the Italian Walk by a series of triumphal arches that accentuated the long prospect like wings; there were the illusionistic prospects, *trompe-l'oeil* scenes at the end of walks, like a painting of the ruins of Palmyra that closed the South Walk (Hunt 1992: 55).

At other times the gardens competed in spectacle with the theatre and Hunt cites the example of a cascading waterfall created at Vauxhall that rivalled that created by Garrick at Drury Lane (Hunt 1992: 56). Water and fountains featured in Renaissance and Baroque gardens and the water theatre, a particular entertainment that amused London crowds from the early eighteenth century, provides another example of the close link between the garden and the theatre (Altick 1978: 77-80).

Hunt argues that not only were gardens designed as theatres but they also 'featured prominently as dramatic locations in intermezzi, operas and plays' (Hunt 1992 : 64). From the masques of Inigo Jones in the early 1600s, who was familiar both with Italian garden as well as stage design, through the design of Restoration scenery and into the eighteenth century, the garden played a pivotal role as *un lieu théâtral* (Hunt 1992: 64-71). As a precursor to the picturesque scene, the spatial formality of the neo-classical garden, with its sharply defined perspectives, suited the scenic constraints of the Restoration and early eighteenth-century stages (Figure 4). And, as Hunt argues, the garden used as a site for deceptions, tricks, ruses and revelations was inherently theatrical so it is unsurprising that a dramaturgy, reliant on these same things, frequently featured scenes set in gardens (Hunt 1992: 67-9).

The revolution in aesthetic theory that occurred in the eighteenth century was inspired, in part, by changing attitudes to landscape. Aesthetic theory prior to the eighteenth century was restricted to the classical dichotomy of Beauty and the Not-Beautiful. In the first decades of the eighteenth century Captain Birt,

7 Again, another major departure from Renaissance uses of *theatrum mundi* where the background 'reality' was always the metaphysical world of heaven and hell. Interpreted by the Church Fathers the 'foolish spectacle' of life used by Roman satirists, such as Lucian, became *contemptus mundi*, as expressed by St John of Chrysostom:

> So then, when we come to the moment of death, having quit the theatre of life, all masks of wealth and poverty will be stripped away—each man will be judged by his works alone: some will be found to be truly wealthy, others will be found poor; some will be honored, others will be scorned (quoted in Christian 1987: 35).

• **Figure 4: Topiary Exhedra with view of the Obelisk Hartwell House, Buckinghamshire, 1738 (Balthasar Nebot, 1738)**
From the Buckinghamshire County Museum collections

a surveyor in the English army of occupation, who surveyed the Scottish highlands in the 1720s, expressed a neo-classical sensibility when, in a letter to a friend, he described the mountains as 'monstrous excrescences ... rude and offensive to the sight ... their huge naked rocks producing the disagreeable appearance of a scabbed head'. In contrast he preferred the beauty of Richmond Hill which was, to his eye, 'a Poetical mountain, smooth and easy of Ascent; cloath'd with a verdant flowery turf where shepherds tend their Flocks' (quoted in Bicknell 1981: ix). Inspired in part by travelling through the Swiss Alps, and in particular their exposure to the art of Salvator Rosa, aristocratic travellers turned to another term to describe the affective power of mountainous landscapes, the magnitude of which overwhelmed them and evoked feelings of horror mixed with delight: this term was the Sublime. Whereas the Sublime suggested wild immensity, vertiginously without limit, and Beauty was associated with cultivated order, smoothness and regularity; in between, was the Picturesque that took on the roughness and irregularity of the Sublime but within human limits. For Gilpin, Beauty and the Sublime were insufficient in themselves: Sublimity without Beauty or Beauty lacking the Sublime had 'little of the picturesque' (Gilpin 1972 [1794]: 43).

In the 1790s the garden became the site for eighteenth-century debates over art and nature, 'improving' or subtly intervening in landscape design according to picturesque principles, out of which emerged the English Landscape Style of gardening. Landscape style proposed that an estate be laid out according to painterly principles with the view from the house organised into a foreground, middle ground and background. The park leading from the house, Gilpin wrote in *Remarks on Forest Scenery*, was 'best displayed on a *varied surface* ... where one part is continually playing in contrast with another' because nature 'seldom passes abruptly from one mode of scenery to another; but generally connects different species of landscape by some third species, which participates in both' (Gilpin 1973 [1791]: 183-184; emphasis in original). The landscape theorists of the 1790s - Uvedale Price, Richard Payne Knight and Humphrey Repton - generally agreed on the picturesque rendering of landscape even when they disagreed, as David Marshall writes, on how this could be achieved (Marshall 2005). The theatricality of picturesque landscaping demanded that gardens be composed with 'pantomimic' elements - ruins of castles and towers, rough hewn bridges, Chinese pagodas and their like - yet this deliberate composition had to remain unrecognised for the effect to work. As Marshall writes, it was 'both the contribution and the curse of the picturesque to inscribe the place of nature in the realm of art' so that although the 'picturesque garden may have aimed to be less theatrical, ... it aimed no less to be theater' (Marshall 2005: 39, 38).

THE SCENE OF THEATRE

It is a curious paradox that the stage so like a picture of the world had not, until the last decade of the eighteenth century, actually managed to integrate the actor into the scene. Indeed, as Christopher Baugh writes, mid-eighteenth century plays still displayed the 'formal qualities of a scenically neutral theatrical place' unlike those produced a century later that relied 'upon pictorial scenography' (Baugh 2007: 43). Playing conventions inherited from the Restoration stage (and earlier) still had the actor performing on a forestage before the curtain line and in front of a decorative scene. With the introduction of moveable wings, shutters and borders, the old discovery scenes could be brought further downstage and deep perspective 'long' scenes - in which the perspective stretched to the back wall of the theatre - became increasingly popular, particularly in opera (Visser 1980: 80, 83). Although the actor did not yet inhabit the stage space behind the proscenium - which was, as Colin Visser writes, 'ultimately inhospitable to the actor, who could not move up or down it without threatening the perspective effect' (Visser 1980: 85) - this space increasingly began to possess environmental qualities that 'invited the presence of the actor and required that he retreat behind the proscenium and define himself in relation to the scenery' (Visser 1980: 89).

The journey of the actor into the scene was assisted by the increasing use of the *scena per anglo* which placed the perspective vanishing point to either side of the stage. This, in turn, allowed the use of set scenes - scenes that involved actual set pieces defined in Rees's 1819 *Cyclopedia* as those items 'which may be occasionally placed and displaced, such as the fronts of cottages, cascades, rocks, bridges, and other appendages' (Dramatic Machinery 1819: n.p.). Not only did the emerging scene involve three-dimensional structures but also the genre of scene, as reflected by Rees's entry, suggests that these set scenes utilised the same scenic elements that were commonly found, as discussed in the previous section, in the compositions of picturesque painting. The spectacle of nature, as I will discuss shortly, would also become a popular subject for English pantomimes in the late eighteenth century.

A key innovator in eighteenth-century scenography, and someone within whom the trajectories of picturesque painting and theatrical practice converged, was Philippe Jacques de Loutherbourg. Loutherbourg worked for the Drury Lane theatre, first for David Garrick from 1772-1776 and then, after Garrick's death, for Richard Sheridan until 1781. David Garrick was, as Baugh argues, the 'supreme delegate' for the eighteenth-century audience; an audience not just for the theatre but for '*all* the product of his age' and his collaboration with Loutherbourg brought together: 'several aspects of shifting audience taste - landscape, topography, the "picturesque", history and the exotic - and the creation of a practical stage machine which could flexibly and speedily show this audience aspects of its own experience' (Baugh 1990: 10; emphasis in original).

Loutherbourg was already a precociously accomplished painter when he arrived in London from Paris with a letter of introduction to Garrick in 1771. He had exhibited at the Salon in 1763, was elected early to the Académie Royale in 1767 and his art had been seen and praised by Diderot (Baugh 1990: 24). His arrival in London, according to Richard Altick, was timely. Garrick had been impressed by the sophistication of Niccolo Servandoni's stage illusions when he had visited Paris some years before and he wished to repeat them in his own theatre. The success of his own pantomimic spectacular, *Harlequin's*

Invasion (1759), encouraged Garrick in his ambition to bring more spectacular effects into his theatre (Altick 1978: 120) and so it was that Loutherbourg worked 'exclusively on spectacles' for Garrick and Sheridan at Drury Lane (Baugh 2007: 47).

What Loutherbourg brought to the English stage was 'what the changing taste at the moment most required, a strong bent for pictorial naturalism'; albeit 'naturalism' with a picturesque tinge (Altick 1978: 119, 120).[8] Loutherbourg, writes Iain McCalman, 'possessed a landscape painter's talent for rendering vivid naturalistic scenes, combined with an engineer's understanding of the mechanics of illusion' (McCalman 2001: 11) and it was he who introduced the *scena per anglo* to the English stage, and experimented with light to produce chiaroscuro effects that had not been seen before (Altick 1978: 120). Despite these innovations Loutherbourg's designs demanded a 'scenographic unity' that was at odds with both the technological capacities of the time and acting conventions which required actors to occupy the forestage and prevented them from being incorporated into the stage picture (Baugh 2007: 47-48).

Perhaps in an attempt to overcome these challenges, Loutherbourg conceived and designed a pantomime that was, as its title suggests, an exemplar of the theatricalised picturesque. *The Wonders of Derbyshire* premiered at Drury Lane in 1779 and although Sheridan was ostensibly the writer for it, he contributed very little to the story, the thinness of which was commented on by the critic in the *St James Chronicle*:

> The subject of the pantomime was judiciously chosen for the display of Mr. Loutherbourg's abilities; but he should have been accompanied into Derbyshire by a Man of some dramatic Genius, or at least of Talents for the Invention of a Pantomime (7-9 January 1779 quoted in Baugh 2007: 48).

As I noted earlier, Loutherbourg had begun touring the Lake District in 1783; prior to this his only other significant excursion into Derbyshire was a tour there in preparation for *The Wonders of Derbyshire* in 1778 (Baugh 1990: 44). Henry Angelo, recalling the settings Loutherbourg created for the pantomime, commented that 'Never was such romantic and picturesque paintings exhibited in the theatre before... [and] gave you an idea of the mountains and waterfalls, most beautifully executed, exhibiting a terrific appearance' (Angelo 1830: 326 and quoted in Baugh 1990: 38).[9] In *The Wonders of Derbyshire* a world that was theatricalised through picturesque conventions of seeing was re-presented in a theatre where those same conventions now appeared. 'Picturesque tourists', accustomed to viewing the world as a spectacle, flocked to Drury Lane where they witnessed a travelogue of similar sights in the theatre which, within its limits, staged the world as a picture.[10] These audiences, as Baugh writes, expected 'to recognize the reality behind the stage image' and 'trust ... the topographical accuracy' of what was portrayed (Baugh 1990: 74). Loutherbourg did not disappoint them.

'Topical spectacles' or pantomimes featuring spectacular representations of current events - military victories, shipwrecks, or picturesque tours of English and 'exotic' locales - were frequently presented in the last decades of the century. Loutherbourg was arguably the greatest exponent of the form and the one most responsible 'for making the eighteenth-century theatre a true image of the form and pressure of its time' (Allen 1965: 289). In 1774 Mai, or Omai as Europeans called him, an islander from Raiatea, travelled to England as a crewman on HMS *Adventure*, a ship that had accompanied Cook on his second Pacific voyage but was forced to return home early.[11] For the next two years, Omai was fêted by English society - meeting the king and queen and having his portrait painted by society painter Joshua Reynolds - and was seen to be the living exemplar of the Noble Savage (Hetherington 2001). Omai's putative 'natural grace' was contrasted with the artificiality of European culture, as Fanny Burney wrote in her diary, Omai had 'an

8 'Naturalism' here must be taken advisedly. Servandoni was a master of magical illusions as was Loutherbourg. The latter's work was more 'life-like' than conventional scenery of the time but not the same as the naturalism of the late nineteenth century.

9 Angelo's adjective 'terrific' was commonly used to describe mountains. The element of fear induced by mountains was a key aspect of the sublime. In a popular letter from John Brown, a friend of the Gilpin family, to George Lyttleton, probably written in 1753 and first published in full in 1767, he describes the lake at Keswick and surrounding countryside which combines the three aesthetic qualities of 'Beauty, Horror, and Immensity'. It is only the painterly skill of Salvator Rosa who would be able to 'dash out the horror of the rugged cliffs, the steeps, the hanging woods, and foaming waterfalls' (quoted in Bicknell 1981: 1-2).

10 Christopher Baugh writes that Loutherbourg's ambition to have the audience 'absorbed' in his parade of wonders would have been embarrassed by 'the illuminated auditorium' which would have mitigated against this (Baugh 2007: 48).

11 Cook undertook three voyages to the South Pacific: 1768-1771; 1772-1775; and his final voyage 1776-1780

understanding far superior to the common race of *us cultivated gentry*' (Cook 2001: 38; emphasis in original). David Garrick even proposed a pantomime in which Omai would become his 'Arlequin Sauvage - a fine character to give our fine folks a genteel dressing' but the pantomime would not eventuate until 1785, at the rival Covent Garden theatre, and by then Omai had been transformed into the role of romantic hero (Cook 2001: 39).

This pantomime, *OMAI, Or A Trip Round the World*, with a script by John O'Keeffe and music by William Shield was designed by Loutherbourg and based loosely on Captain Cook's first South Pacific voyage over ten years earlier, recorded in an edited collection by John Hawkesworth in 1773.[12] *OMAI* was, according to Iain McCalman, 'the greatest blockbuster pantomime of the eighteenth century' featuring nineteen spectacular scenes, designed by Loutherbourg and based on the paintings by John Webber who accompanied Cook on his last voyage (McCalman 2001: 10, 11-12).[13] On the one hand these images of Polynesia reflected an eighteenth-century empiricist consciousness that sought to accurately represent the world as it appeared to the observer. Yet, on the other hand, like Henry Baker with his microscopic magic lantern shows, fascination with the sheer strangeness of these 'other' worlds impelled a showman's response to them. Thus the review in *The Rambler* considered *OMAI* to be 'an illustration of importance to the mature mind of an adult, and delightful to the tender capacity of an infant'; furthermore, it continued:

> To the rational mind what can more entertaining than to contemplate prospects of countries in their natural colours and tints - to bring into living action, the customs and manners of distant nations! To see exact representations of their buildings, marine vessels, arms, manufactures, sacrifices and dresses? (January 1786, quoted in McCalman 2001: 11).

Yet the illustrations by John Webber and William Hodges were themselves mediated representations of reality and McCalman observes that Loutherbourg treated the plates 'as a rich menu of possibilities' for theatrical rather than 'ethnographic' representation (McCalman 2001: 12). Bernard Smith argues that artists such as Hodges and Webber, and certainly the engravers employed by Hawkesworth, Giovanni Battista Cipriani and Francesco Bartolozzi, were all schooled in neo-classical conventions of history painting: this accounts, he suggests, for the idealised images of Pacific Islanders that appeared both in Hawkesworth's *Account* and Loutherbourg's designs for *OMAI*, among numerous others (Smith 1992: 60-61). In any case, these artistic renderings of the alien landscapes and people encountered by the Europeans in the South Pacific were rendered 'knowable' through conventions of the picturesque and history painting.

CONCLUSION

The two hundred year old theatrical metaphor encouraged spectators in the eighteenth century to view the world as if it were a stage-set upon which humanity played and watched itself play. This world scene was no neutral backdrop but was itself a source of wonder; moreover, it could evoke strong emotions in the spectator. Even when the theatre of the day was not quite up to the job of pictorially representing the world; nonetheless, in the way it spatially organised figures and scenes, and how it made the spectacle of nature spectacular, it reinforced the idea of theatre in the *theatrum mundi*. *Theatrum mundi* expressed an ontological sense of place through its structural relationships of player/place/spectator, and through its dramaturgy of players and roles. Although the theatrical metaphor suggested a particular spectatorial positioning, to picturesque painters and gardeners, the idea of the world stage no longer had the theological and humanist resonances of *theatrum mundi*. It emphasised instead the spectacular visuality of the world, but a world that was now safely bounded as a 'view'.

However, it was theatricality as an idea, rather than the theatre itself, that influenced how

[12] Hawkesworth, John (1773) *An Account of the Voyages Undertaken by Order of His Present Majesty, for Making Discoveries in the Southern Hemisphere, and Successively Performed by Commodore Byron, Captain Wallis, Captain Carteret, and Captain Cook, in the Dolphin, the Swallow, and the Endeavour: Drawn up from Journals Which were Kept by the Several Commanders and from the Papers of Joseph Banks, Esq.*, London: W. Strachan and T Cadell.

[13] Webber also was credited as one of the scene painters for *OMAI*. The artists who accompanied Cook on his earlier voyages were: Sydney Parkinson, Alexander Buchan and Herman Spöring on the first voyage; William Hodges and George Forster on the second (some of Hodges' paintings were also referenced for *OMAI*).

spectators in the eighteenth century viewed the worlds that appeared to them: both microscopically magnified and worlds enlarged through exploration. Theatricality was, if you like, the *Zeitgeist* of the age and theatre had become, as Baugh, writes 'an available metaphor - a vehicle for the ordering, structuring and criticism of life' (Baugh 1990: 19). For Gilpin and others, who sought an affective representation of the world through painting, theatrical seeing enabled them to put the world they saw into the frame of art. The painters on Cook's voyages - and even more so the engravers that reproduced their work - represented the peoples of the South Pacific in a series of scenes which, for Loutherbourg, could then easily be rendered in the theatre. The artistic frame, in this way, became re-theatricalised.

There is, at the heart of the term 'picturesque world stage', a paradox that this essay has attempted to unpack. The late eighteenth-century spectator demanded greater accuracy in the theatrical and artistic representation of the world in keeping with the empiricist ideals of the Enlightenment. Yet, the ontology of the theatrical metaphor suggested that the world that appeared to these spectators was a spectacle placed there to entertain them. Just as a scene played in a theatre could delight so too could views framed as scenes by the 'sensitive' traveller provide similar pleasures. Here the conventions of the picturesque outlined by Gilpin, and argued about and satirised by others, required an artistic rendering of the world at odds with naturalism because pictorial representation of the world depended on theatricalised seeing. No longer was the world a bare platform upon which the 'poor strolling player' strutted his stuff but was now an artfully composed picture engaged in a struggle with the terms of its own representation.

REFERENCES

Allen, Ralph G. (1965) 'Topical Scenes for Pantomime', *Educational Theatre Journal* 17 (4): 289-300.

Altick, Richard Daniel (1978) *The shows of London*, Cambridge, Mass.: Belknap Press.

Angelo, Henry (1830) *Reminiscences of Henry Angelo*, 2 vols., Vol. 2, London: Colburn & Bentley.

Balme, Christopher (2006) Metaphors of Spectacle: Theatricality, Perception and Performative Encounters

• *(left)* **Figure 5: The present woman of Oteheite (Philippe Jacques de Loutherbourg, 1785) National Library of Australia, Canberra** *(http://nla.gov.au/nla.pic-an2668156)*

• *(right)* **Figure 6: A young woman of Otaheite, bringing a present. (J. Webber del, F. Bartolozzi sc., 1784) National Library of Australia, Canberra** *(http://nla.gov.au/nla.aus-nk1428-4-s27)*

in the Pacific metaphorik.de, http://www.metaphorik.de/aufsaetze/balme-theatricality.htm#_ftn6

Baugh, Christopher (1990) *Garrick and Loutherbourg.* Edited by R. A. Cave, *Theatre in Focus*, Cambridge and Alexandria, VA: Chadwyck-Healey.

Baugh, Christopher (2007) 'Scenography and technology' in J. Moody and D. O'Quinn (eds) *The Cambridge Companion to British Theatre, 1730–1830*, Cambridge: Cambridge University Press.

Bicknell, Peter (1981) *Beauty, Horror and Immensity. Picturesque Landscape in Britain, 1750–1850*, Fitzwilliam Museum, Cambridge: Cambridge University Press.

Christian, Lynda G (1987) Theatrum mundi: the history of an idea, Thesis (Ph. D.), Garland, Harvard University, 1969, New York.

Cook, Alexander (2001) 'The Art of Ventriloquism: European Imagination and the Pacific' in M. Hetherington, I. McCalman and A. Cook (eds) *Cook and Omai: the Cult of the South Seas*, Parkes, ACT: National Library of Australia.

Curtius, Ernst Robert (1953) *European literature and the Latin Middle Ages*, trans. W. R. Trask, London: Routledge and Kegan Paul.

Dramatic Machinery (1819) *The Cyclopedia; or, Universal Dictionary of Arts, Sciences, and Literature*, ed. A. Rees, London: Longman, Hurst, Rees, Orme, and Brown.

Esau, Erika (1991) 'Tiepolo and Punchinello: Venice, Magic and Commedia Dell 'Arte', *Australian Journal of Art* IX (Theatricality): 41–57.

Fried, Michael (1980) *Absorption and Theatricality. Painting and Beholder in the Age of Diderot*, Chicago and London: University of Chicago Press.

Gilpin, William 1972 [1794]. *Three essays: On picturesque beauty; On picturesque travel; and On sketching landscape: to which is added a poem On landscape painting*, 2nd ed., Farnborough, Eng.: Gregg.

Gilpin, William 1973 [1782]. *Observations on the River Wye*, 1st ed., Richmond: The Richmond Publishing Co. Ltd.

Gilpin, William 1973 [1791]. *Remarks on Forest Scenery*, Richmond: The Richmond Publishing Co. Ltd.

Heidegger, Martin (1977) *The question concerning technology and other essays*. Paperback ed., *Harper torchbooks*; TB 1969, New York: Harper & Row.

Hetherington, Michelle (2008) *Cook and Omai: the Cult of the South Seas* [online catalogue], National Library of Australia 2001 [cited 28 October 2008], available from http://www.nla.gov.au/exhibitions/omai/.

Hipple, Walter John (1957) *The beautiful, the sublime and the picturesque in 18th century British aesthetic theory*, Carbondale, Ill.: Southern Illinois University.

Hunt, John Dixon (1992) *Gardens and the picturesque: studies in the history of landscape architecture*, Cambridge, Mass.: MIT Press.

Hussey, Christopher (1967) *The Picturesque. Studies in a Point of View*, London: Frank Cass and Co. Ltd.

Marshall, David (1986) *The Figure of Theater. Shaftsbury, Defoe, Adam Smith and Geoge Eliot*, New York: Columbia University Press.

Marshall, David (2005) *The Frame of Art. Fictions of Aesthetic Experience, 1750–1815*. Edited by G. P. Stephen, G. Nichols and Wendy Steiner, *Parallax Re-Visions of Society and Culture*, Baltimore: The Johns Hopkins University Press.

McCalman, Iain (2001) 'Spectacles of Knowledge: OMAI as Ethnographic Travelogue' in M. Hetherington, I. McCalman and A. Cook (eds) *Cook and Omai: the Cult of the South Seas*, Parkes, ACT: National Library of Australia.

Pike, Joseph B., ed. (1938). *Frivolities of Courtiers and Footprints of Philosophers: Being a Translation of the First, Second, and Third Books and Selections from the Seventh and Eighth Books of the Policraticus of John of Salisbury*, Minneapolis: The University of Minnesota Press.

Smith, Bernard (1992) *Imagining the Pacific in the Wake of the Cook voyages*. Carlton, Vic.: Melbourne University Press at the Miegunyah Press.

Stafford, Barbara Maria (1993) Voyeur or Observer?: Enlightenment Thoughts on the Dilemmas of Display, *Configurations* 1 (1): 95–128.

Stafford, Barbara Maria, Isotta Poggi, Frances Terpak, and J. Paul Getty Museum (2001) *Devices of wonder: from the world in a box to images on a screen*, Los Angeles, CA: Getty Research Institute.

Steele, Richard (1720) *The Theatre* No.7 (23 January): 51–59.

Visser, Colin (1980) 'Scenery and Technical Design' in R. D. Hume (ed.) *The London Theatre World 1660–1800*, Carbondale and Edwardsville: Southern Illinois University Press.

Sweating Blood
Intangible heritage and reclaimed labour in Caribbean New Orleans

JOSEPH ROACH

Until recently, not many people outside of Louisiana ever thought of New Orleans as an island. But it is. Bounded by the Mississippi River, lakes, and wetlands, the Big Easy waited until the 1880s, when the railroads came through on raised embankments and long tressells, for a way to travel to or from the city by any means other than ship or boat. Assessing its place in a regional interculture of islands, we might think of New Orleans not only as an island, but as a 'repeating island' of the kind described by Antonio Benítez-Rojo (1992: 3) - an island like Jamaica, Haiti, or Cuba - each repeating a colonial and post-emancipation history in its distinctive way, and each asserting its geo-historical place as part of a bridge of islands connecting America, North and South. Benítez-Rojo calls this a 'meta-archipelago' (4), about which we no longer say that a New World was discovered in the Caribbean five hundred years ago, as many of us learned in school, but rather that a new world was invented there. In the face of the clear and present danger of so-called natural disasters - weather events exacerbated by environmental degradation, engineering folly, and political negligence - performance researchers need to come to a better understanding of how this fragile but fecund world continues to be re-invented in New Orleans and elsewhere, from Trinidad to Toronto, despite (and because of) its vulnerabilities. The resiliency it has shown, for expatriates and tourists as well as natives, is founded on its tenaciously regenerative performances, the compelling playfulness of which, I will argue, derives from their peculiar historical relationship to work. Work defines many of the experiences that the islands share, especially play.

The first and pre-eminent among these is one that has not been properly acknowledged by name, in part because it isn't very polite to talk about it in formal settings. It must be emphasised nevertheless, because it unites the past and present experience of the archipelago: sweat. The Irish call the North Atlantic a bowl of tears. Let the Caribbean basin, after centuries of forced labour in tropical and semi-tropical climates, be re-christened the sea of sweat. This is particularly appropriate to those preternaturally humid places, which, like Louisiana or Jamaica, produced sugar, the most labour intensive of all agricultural enterprises. But these sweat zones, regional intercultures of excessive perspiration, also produced the most ambitiously labour-intensive festivals anywhere. And they continue to do so, in season or out, because carnival remains not only the most beloved local heritage and preferred state of mind, but also the best cash crop.

The tourism-boosting attraction of similarly labour-intensive festivity recently reasserted itself in the run-up to the 'Prospect.1' Biennial, the ambitious exhibition of international art that opened in New Orleans in November 2008. This circumspect Mardi Gras for high-end cultural consumers appeared in venues throughout the Crescent City, a number of them reclaimed from

Performance Research 13(4), pp.140-148 © Taylor & Francis Ltd 2008
DOI: 10.1080/13528160902875721

flood-damaged buildings, featuring works by eighty one artists from thirty four countries. Reclamation was a principal technique as well as a major theme of P.1. 'The Ark', an outdoor construct by Los Angeles-based artist Mark Bradford, for instance, epitomised the bricolage of the endeavor, making work by making do in the heavily damaged Lower Ninth Ward. The hull and superstructure of the 30-foot tall vessel rose improbably above the shells of vacant houses and overgrown lots. Built of salvaged sheets of three-quarter plywood, which is the material par excellence to post-Katrina New Orleans as travertine is to Tivoli, the Ark seemed to express more determination than irony. It served as a symbolic point of disembarkation for the brave souls - now more than a handful, still less than a neighborhood - who have returned and are rebuilding their homes. The local organisers worked hard to make good on P.1's allegory of starting anew after the Flood. The depth and resurgent power of the culture that their labour reclaims appeared all over the city, inside and outside the designated venues, but most poignantly in the high-culture sanctum sanctorum of the New Orleans Museum of Art (NOMA). At NOMA, P.1 sponsored an exhibition of the heart-stoppingly beautiful, hand-sewn 'suits' of Mardi Gras Indian Victor Harris, Big Chief of the Fi Yi Yi and Mandingo Warriors. Over the past thirty years, Harris, along with Allison 'Tootie' Montana of the Yellow Pochahontas, Larry Bannock of the Golden Star Hunters, and Bo Dollis of Wild Magnolias, led a brilliant ensemble of Mardi Gras Indian 'gangs'. Their network of putative affiliation dates back to the bands of Maroons who defied slavery in colonial times. Identifying with Native Americans in their post-genocidal reclamation of a threatened cultural heritage, contemporary Mardi Gras Indians are supremely accomplished craftspeople and performers. In many places in the world today they would be designated 'Living National Treasures', in the manner of the great Japanese artists of the kabuki and Noh stages; but in New Orleans they risk being busted for parading without permits. Like the city itself, the heritage that the Mardi Gras Indians have repossessed is under threat. But it is also under consideration as a cultural alternative in other North American cities, large and small, where masqueraders transplanted from the repeating islands seize the opportunity of their displacement to show the local workers how to play hard.

Two recent works of scholarship feature Caribbean performance history to show how truly renovated an idea 'America' might be. The first is the anthology entitled *Just Below South: Intercultural Performance in the Caribbean and the US South*, edited by Michael Bibler, Jessica Adams, and Cècille Accilien. Published by the University of Virginia Press in an act of commendable self-interrogation (post-Katrina) for all that it found itself doing to honour the four hundredth anniversary of the founding of Jamestown Colony in 1607, *Just Below South* proposes a basic geo-historical reorientation of how the region is to be understood in terms of the deeply historic but newly acknowledged meta-archipelago . Standing at Jamestown and looking west, visitors imagine the birth of one kind of America, the destiny of which was once thought to be manifest. But standing in New Orleans and looking around, the prescient editors imagine many more. These Americas - vibrantly emergent but sometimes still invisible, historically vulnerable but even now filled with possibility for the future - are explored by eleven specialist contributors in literature, festival arts, history, dance, linguistics, and history. They map a region of signs and wonders that stretches from Tobago to Tennessee.

The second publication is the catalogue and exhibit organised by the Paul Mellon Center for British Art at Yale in 2007, called *Art & Emancipation in Jamaica: Isaac Mendes Belisario and his Worlds*, curated by Tim Barringer, Gillian Forrester, and Barbaro Martinez-Ruiz. Drawing on the research of art historian Dian Kriz of Brown University, the Belisario show brought new perspectives to bear, not only on the subject of British art and culture as heretofore received at the British Art Center, but also on the manifold ways whereby a repeating island like Jamaica can disclose to North Americans a world they ought to have known better all along while prophesying one that might be closer within reach than ever before - a world in which a Caribbeanised diversity of peoples and art shows us the way to a more culturally festive global future. The arguments and images of *Art and Emancipation in Jamaica* resonated in the works of the P.1 Biennial, especially in the hosanna of beads and feathers of the Mardi Gras Indian suits at NOMA. In both New Haven and New Orleans, however, the festival costumes hung motionless on mannequins, out of the cultural context of their function in the liveliest of street performances. The knowledgeable and imaginative curators offset the stiltedness of the exhibits with video supplements and, in New Haven, with supplementary performances by dancers and musicians. Such demonstrations, viewed only as spectacles, felt somehow incomplete, because they summoned witnesses rather than participants. But when the spectators began to sway with the music and add to the counter-rhythmic accompaniment, 'working it out', with the variety of percussion instruments provided, they could make an active contribution, however small, to the tradition they had come to consume. Otherwise, their predicament was the predicament of heritage itself, which retains its true vitality only for the ones working the hardest to keep it alive.

In his keynote for the Belisario show's opening, Rex Nettleford, Vice Chancellor of the University of the West Indies, spoke movingly about the urgent but troubling concept of 'Intangible Heritage'. Professor Nettleford referenced an initiative by the Cultural Heritage Division of the United Nations Educational, Scientific and Cultural Organization (UNESCO), which designates world-historic cultural performances as worthy of preservation. Intangible heritage exists not as tangible objects or permanent structures, but as embodied practices, uniquely alive and yet repeatable as behaviors. The umbrella structure for this impressive and impressively problematic enterprise is called 'World Heritage'. The slogan of the program betrays its well-intentioned but contradictory agenda: 'Cultural Diversity: A New Universal Ethic'. The basic idea is that some things made on Planet Earth are too important to belong solely to the people who created them: their preservation is important to all humankind.

Although the elegiac language of salvage ethnography - catch it before it disappears forever - is made more plausible by practical recommendations for preservation (often by promoting it as a tourist attraction), the impulse is not new to the repeating islands. The nostalgic images reproduced in Isaac Mendes Belisario's *Sketches of Character: in Illustration of the Habits, Occupation, and Costume of the Negro Population in the Island of Jamaica* mark it as an early precursor to more recent attempts to identify and fix 'intangible heritage' before it slips away. Belisario recognises the changes that impinge on the traditional performances even as they happen before his eyes. But the characteristic of 'Live Arts' is that they can't be fixed. They adapt and grow as they pass like a wave through the three great constituencies of art and culture - the three constituencies identified independently but in almost the same words by the otherwise unlikely pair of Edmund Burke, the eighteenth-century political philosopher, and Wole Soyinka, the Nigerian Nobel Laureate - the three great constituencies being: the dead, the living, and the yet-to-be-born.

One of the early models for the Intangible Cultural Heritage initiative was the Japanese policy of identifying exceptional practitioners of traditional arts and crafts as 'Living National Treasures'. Another early strategy was to identify 'Masterpieces of Intangible Heritage' world-wide. As the number of designated Intangible Heritage sites proliferates, these categories have been interpreted to include everything from culinary arts to the subjunctive in a language, as well as entire languages themselves. Even entire extended communities are eligible for designation as World Intangible Heritage sites. Moore Town in Eastern Jamaica, for instance, is now recognised under UNESCO protocols as an Intangible Heritage site for its living tradition of Kromanti Play. These spirit-world ritual practices, which trace their origins to the rituals of Jamaican Maroons, constitute what the Cultural Division of UNESCO calls 'a world heritage masterpiece of oral tradition'. The Maroons, bands of escaped slaves who lived autonomously (and adversarially) in their own communities in the Blue and Johncrow Mountains of Jamaica, kept the British colonial authorities at bay for generations, developing a unique and powerfully self-sustaining tradition of live arts in the service of both the sacred and social, affirming their solidarity in the face of determined oppressors and long odds.

In the fecundating sea of sweat, the Jamaican Maroons did not act alone, but in symbolic and actual connection with other, parallel traditions. Both Trinidad and New Orleans, for instance, at the furthest edges of the region, host renegade 'Black Indians' as carnival favorites. Many participants are struck by the closeness of the practices of Trinidadian carnival and Mardi Gras in New Orleans. So much so that Frederick Street's lithograph of Melton Pryor's sketch for *The Illustrated London News* of 1888 has been reproduced a number of times as a scene from Mardi Gras on Bourbon Street in the French Quarter of New Orleans, with its narrow streets, its iron balconies, and its flamboyantly crapulent masquerades, in costumes that aspire to ambulant architecture. But it is not New Orleans. Street and Pryor are depicting Port of Spain, Trinidad, as the original caption shows.

No one would deny that there are many and important cultural differences between these two festival traditions at the far opposite ends of the Caribbean and the Gulf of Mexico - Trinidad and Tobago and New Orleans, Louisiana - and more differences still between each of them and Jamaica. But there are some striking similarities, founded on the heritage of work and play in the copious expenditure of sweat.

• **Plate 1. Carnival in Port of Spain, Trinidad.** *Author's collection*

• Plate 2. Koo, Koo, or Actor Bo, I. M. Belisario. *Yale Center for British Art, Paul Mellon Collection.*

Jamaican Jonkonnu is a Christmas revel, but the subversive laws of carnival inversion and pre-Lenten subversion still obtain. The first of Belisario's two lithographs from *Sketches of Character* depicting 'Koo Koo, or Actor Boy', the one showing a Kingston street scene dated 1837, has several features in common with the 1888 version of carnival in Port au Prince published by *The Illustrated London News*. The costume again functions as ambulant architecture. With broad gestures and protruding papier-mache prostheses topped off with feathers, the masker claims the space through which he dances. The very large drum that accompanies him must be imagined to extend his momentarily privileged sway sonically over the entire neighborhood. His mask is whiteface - silly, immobile, and ironic - set within the blaze of multi-coloured fabrics and beads and set in motion by the expressive gestures of his dancing body. Belisario leaves it for the reader to imagine the dissonant effect of the perpetually smiling mask on the changing moods of the Actor Boy's pantomime.

• Plate 3. Larry Bannock, Big Chief of the Golden Star Hunters. *Image by Michael P. Smith. © The Historic New Orleans Collection*

The Mardi Gras Indians of New Orleans parade during carnival season and on special days during Lent on unannounced routes. They appear in colourfully elaborated suits, beads and feathers sewn onto multi-layered underskirts and heavily decorated tabs. In basic silhouette they resemble the Actor Boys' layered-look regalia, but Mardi Gras Indians, with the notable exception of Victor Harris, customarily parade unmasked. As in Jamaica of old, the Mardi Gras Indian suits take many months, sometimes a whole year, to construct. Traditionally, at the end of the masking season, the suits are destroyed, reverently cut up with seam-rippers, and the process begins all over again, always in competitive pursuit of being judged the 'Prettiest' Indian at the next year's carnival. This is Larry Bannock, Big Chief of the Golden Start Hunters, pictured on Mardi Gras morning near his home in Gert Town in 1983. Like all the Mardi Gras Indian Chiefs, Larry Bannock had to work a full-time day job, in his case as cab driver, because beads and feathers are expensive, even if he and his fellow gang members contribute their labour in sewing their suits. The photo is by Michael Smith of New Orleans, an artist in his own right and recorder of the secret carnivals of the 1970s, 80s, and 90s. The laborious process of sewing the suit requires fixing each bead on a thick canvas backing, using an upholstery needle for the purpose. It is hard work to do. As Mardi Gras approaches, the calloused hands of the chiefs begin to blister and bleed, a part of the ritual. Larry Bannock once told me that he 'sweats blood' for each suit. He believes that his heritage as a Mardi Gras Indian Chief originated with the Louisiana Maroons, escaped slaves who lived with Native Americans in freedom on the swampy peripheries of New Orleans. He always answered my questions patiently and generously, except once, when I asked him why he was willing to sweat blood for Mardi Gras night after night, day after day, between long shifts in the cab. Taking my measure slowly and shaking his head as if I must be incapable of grasping the obvious and most important truth, he replied, 'Because it's fun, Fool.'

• **Plate 4. Koo, Koo, or Actor Bo, I. M. Belisario.** *Yale Center for British Art, Paul Mellon Collection.*

Typically, every part of the body of the Mardi Gras Indian is covered by sequins and feathers and ornaments of daring scope and invention – except for his face. That suggests to me that in the secret carnivals of New Orleans, at least, and perhaps elsewhere in the repeating islands, there is a mask that is worn in everyday life, sweating for a living. The true face comes out only at carnival time, sweating for a life. Why is that? Why do high-energy festival arts flourish in the interstices of such labour-exhaustive conditions? Could it be as simple as this: Those who work the hardest, play the hardest? I am proposing that the conditions of alienated labour – unremunerated before emancipation, scarcely remunerated after it under various forms of a vassalage and exploitation – impelled another form of expenditure in the ritualised occasions of leisure. This one was based on publicly exhibited effort through performance – Jonkonnu, carnival, mas, Mardi Gras – which characteristically in the

• Plate 5. New Orleans in Mardi Gras Time. *Author's collection*

repeating islands take a year's work prior to the few red-letter days of sacrificial expenditure. Such a fevered release of energy expresses, literally and figuratively, the sweat that the performers have invested on their own time, doing their own work, which may then be experienced as a kind of free play. In the logic of such playful expression, would not sweat, covering the face, then become a mask? Would not that perforce be a festive mask, secretly (and secretively) expressing time and labour taken back from those who stole them in the first place? In Africa and in Afro-diasporic cultures, the mask puts the wearer in touch with the magic. (Plate 4. Belisario's Koo Koo, Actor Boy #2.) Belisario's second version of 'Koo Koo the Actor Boy' from *Sketches of Character* shows him lifting the white mask to reveal his face, allowing him to use his ornamental fan better to cool his sweating brow. But the playful smile under the fixed profile of the tipped mask suggests a knowing wink as well.

The struggle for the space to work and play freely, meeting your interlocutor eye to eye, on even terms, is part of the tortured history of carnival in the shadow of slavery, not only in the period of slavery, of its 'world', but during the following period of its transmutation. This post-emancipation moment of transition is captured poignantly in the Belisario images, as a world between two worlds, one dead, as Matthew Arnold put it, the other still powerless to be born. It is also captured in images of the New Orleans carnival during the period of post Civil War Reconstruction, which was quickly and violently overthrown by a coup d'etat in 1874, and the successor period of Bourbon Redemption in Louisiana, a 'world' all its own and unto itself. The parading traditions of the all-white Mardi Gras 'krewes' of New Orleans required that the floats be accompanied at night by flambeaux, African Americans carrying the torches and managing the mule train that pulled the wagons. The spectacle of the festival parade, with masqueraders taking their leisure on the floats, was incomplete without sweating African-American faces, glistening in the torchlight, here shown in Joseph Pennell's lithograph, 'In Carnival Time - New Orleans, 1884'.

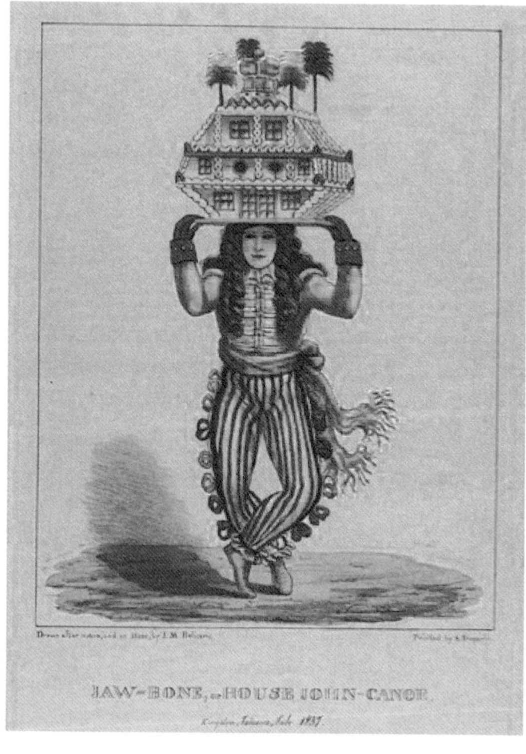

• Plate 6. Jaw-Bone, or House John-Canoe. I. M. Belisario. *Yale Center for British Art, Paul Mellon Collection.*

• **Plate 7.** *Image by Michael P. Smith. © The Historic New Orleans Collection*

So the question imposed by carnival on its busiest participants was and is, for what, or for whom, do you sweat? The fantastic elaboration of carnival costumes and festival performance in the African-diasporic traditions considered saliently here suggests that there is a deeper motivation than carnivalesque inversion of the elite power structure, although the satisfactions of that are self-evident. The man with the house on his head is the central figure in the festivity that gives its name to the paste-board palace, with its colourful ornaments and improvements, here rendered by Belisario in *Sketches of Character*. The house might be an attractive place for spirits to reside and refresh themselves during festival time. Food can be left in there to add to the inducements. Jaw-Bone, the man who carries the house on his head, is a youthful dancer with a lively step. This is the Jamaica of Isaac Mendes Belisario.

By way of comparison across the meta-archipelago, Michael Smith's 1981 photograph shows a young Mardi Gras Indian masker on Orleans Avenue on Super Sunday, a special day of parading during Lent in New Orleans. (Plate 7. Child Mardi Gras Indian Masker.) The reason that Intangible Heritage of the repeating islands may be preserved is that its forms are communicable through embodied practices - not just the letter but the spirit. The trace of Intangible Heritage is left by the record of the sweat of the performers, appearing as the mask handed down to cover the faces of those next in line. (Plate 8. Comparison of Child Mardi Indian and Jaw Bone, House of John-Canoe.) The dancer-born house is perhaps a space to attract and welcome friendly spirits hospitably, and the dynamic kinesthetic movements might reflect the powers infused by such a visitation by the ancestors. With musical effects and choreography, the architecturally ambulant costume claims the space through which it moves, with the choric followers, the 'Second Line' in New Orleanian terms, consolidating the space that the passage of the sacred procession is clearing before them, like succeeding generations of the dead, the living, and the yet-to-be born. In that space, change must come about, even as tradition reigns. In that space, something old and yet new must appear, something truly worth sweating for.

Hurricane Katrina flooded eighty per cent of New Orleans and left sixty per cent of the houses ruined beyond repair. Almost every Mardi Gras Indian lost his or her home. Many lost a life's work in collected beadwork and photo records. Some have left New Orleans, never to return. Others have returned only to leave again. The ones who have come back to stay labour mightily to reclaim their damaged hopes as well as their property. But their artistic ways and means of reclaiming a violently threatened heritage are strong - tempered by a legacy of calamities followed by the motivating incentive of festively reclaimed labour. The spirit of art as reclamation visited the houses built for its reception in the multiple venues of Prospect.1 Biennial. At the Charles J. Colton School, for instance, the classrooms, dining hall, and auditorium have been converted to artists' studios and exhibition-installation spaces. Like the man with a house on his head, Colton School bids the departed to return to carry on the work of the repeating islands. Enough of them did in 2008 to boost tourism. Heritage is the name that UNESCO gives to the beauty of their labour. They call it a way of life. At its heart, their idea is potentially emancipatory for many of the rest of us, a leap from the work we do only because we must, which is a dreary kind of slavery, to the work that we do because, as Larry Bannock said, 'It's fun, Fool,' which is an exhilarating kind of freedom.

REFERENCES

Barringer, Tim J., Gillian Forrester, and Barbaro Martinez-Ruiz (2007) *Art & Emancipation in Jamaica: Isaac Mendes Belisario and his Worlds*, New Haven: Yale Center for British Art, Yale University Press.

Belisario, Isaac Mendes (1837–38) *Sketches of Character: in Illustration of the Habits, Occupation, and Costume of the Negro Population, in the Island of Jamaica*, Kingston: Published by the Artist.

Benitez-Rojo, Antonio (1992) *Repeating Island: the Caribbean and the Postmodern Perspective*, trans. James Maraniss. Durham: Duke University Press.

Bibler, Michael, Jessica Adams, and Cécile Accilien (2007) *Just Below South: Intercultural Performance in the Caribbean and the US South*. Charlottesville: University of Virginia Press.

Kriz, Kay Dian (2008) *Slavery, Sugar, and the Culture of Refinement: Picturing the British West Indies, 1700–1840* New Haven: Yale University Press for the Paul Mellon Centre for Studies in British Art.

Nunley, John W. and Judith Bettleheim (1988) *Caribbean Festival Arts: Each and Every Bit of Difference*, Saint Louis: Saint Louis Art Museum with the University of Washington Press.

Smith, Michael P. (1994) *Mardi Gras Indians*, Gretna, La.: Pelican Press.

Performance Research: On Appearance

Notes on Contributors

THE EDITORS

Ric Allsopp is a founding editor of *Performance Research*. He is currently Senior Research Fellow in the Department of Contemporary Arts at Manchester Metropolitan University, UK, and Visiting Professor at the University of the Arts, Berlin.

Richard Gough is general editor of *Performance Research*, a Professor at Aberystwyth University, Wales, and Artistic Director for the Centre for Performance Research (CPR). He has curated and organized numerous conference and workshop events over the last 30 years as well as directing and lecturing internationally.

ISSUE CO-EDITOR

Adrian Kear is Professor of Theatre and Performance and Head of the Department of Theatre, Film and Television Studies, Aberystwyth University. He is the co-editor of *Psychoanalysis and Performance* (Routledge 2001) and *Mourning Diana: Nation, Culture and the Performance of Grief* (Routledge 1999). He contributes to the academic journals *Contemporary Theatre Review*, *Parallax* and *Moving Worlds* as well as frequently writing for *Performance Research*. He is currently completing a book entitled *Theatre and Event* (forthcoming 2009).

CONTRIBUTORS

Fêted for his astonishing sleight-of-hand conjuring prowess and for his rare work in interdisciplinary culture, legendary magician **aladin** separately has an international reputation as a strategy and organisational development consultant and thinker. Based in London, aladin's practice encompasses a vast array of interventions across civil society. www.aladin.me www.aladinmagic.com www.subvirt.com

Richard Allen is a PhD candidate at Aberystwyth University working towards a practice-led thesis on 'The Object as Postdramatic Gesture'. His animated object theatre has most recently been performed at Chapter Arts Centre (Cardiff), The National Review of Live Art (Glasgow) and the Central School of Speech and Drama (London).

Simon Bayly is a Principal Lecturer in Drama, Theatre and Performance Studies at Roehampton University, London, and an artist and writer working broadly in the fields of performance, primarily with the London-based collective PUR.

Kasia Coleman is a visual artist and theatre designer primarily working with drawing as the impetus for making performance. The Galapagos Man, her first collaboration with Richard Allen, was invited to the National Review of Live Art in February 2008. She is currently working on a performative drawing project at Aberystwyth University.

Gerald Davies is an artist and Senior Lecturer in Drawing at the Lancaster Institute for the Contemporary Arts (LICA). Over the years he has walked - notebook in hand - the rafters of Durham Cathedral, examined body parts in a teaching hospital, drawn medieval plague pits and recently drawn deep in the caves of North Yorkshire. The commonality is to connect a moment or fragment of social or cultural history with a vivid and timely image or experience to form a new work with contemporary relevance.

Jim Drobnick is a critic, curator and Associate Professor of Contemporary Art and Theory at the Ontario College of Art and Design, Toronto. He has published on the visual arts, performance and post-media practices in anthologies including *Crime and Ornament* (2002) and *Empire of the Senses* (2004) and in the journals *Angelaki*, *High Performance*, *Parachute*, *Performance Research* and *The Senses and Society*, where he is now reviews editor. His books include the anthologies *Aural Cultures* (2004) and *The Smell Culture Reader* (2006).

Musetta Durkee received an MA in Performance Studies from New York University (2007) following a BA in philosophy from Columbia University (2006). She is currently an independent scholar and writer addressing topics from human rights and citizenship to performance and the visual arts.

Kathleen Gough is a lecturer in the Theatre, Film and Television Studies Department at the University of Glasgow where she teaches courses in intercultural performance and postcolonial drama. She is currently writing a monograph entitled *Gender and Performance in the Black and Green Atlantic*.

Geraldine (Gerry) Harris is Professor of Theatre Studies, Lancaster University. Recent books include *Beyond Representation: The Politics and Aesthetics of Television Drama* (Manchester University Press 2006) *Feminist Futures? Theatre, Theory, Performance*, co-edited with Elaine Aston (Palgrave 2006) and *Practice and Process: Contemporary [Women] Practitioners*, co-authored with Elaine Aston (Palgrave 2007).

Performance Research 13(4), pp.149-150
DOI: 10.1080/13528160902875739

Joe Kelleher is Professor of Theatre and Drama and Head of Subject: Drama, Theatre and Performance in the School of Arts at Roehampton University, London. He is the author of the forthcoming book *Theatre and Politics* (Palgrave Macmillan 2009).

Kinkaleri was founded in 1995 as a grouping of formats and means, balancing in the attempt. From the outset the company has worked in a number of different directions and areas: plays, performance pieces, installations, videos, soundtracks, publications. Kinkaleri is currently based at the Spazio K in Prato, Italy. www.kinkaleri.it

Anthony Kubiak is Professor of Drama at the University of California, Irvine. His most recent book, *Agitated States: Performance in the American Theater of Cruelty*, investigates the uses of theater and theatricality in the stagings of American history. He is also the author of *Stages of Terror: Terrorism, Ideology and Coercion as Theatre History*. His current project - a series of meditations on art, spirit and the unconscious - is provisionally titled *Ecopoesis: Art as Survival.*

Carl Lavery teaches theatre and performance at Aberystwyth University. He has edited (with Clare Finburgh and Maria Shevtsova) *Jean Genet: Performance and Politics* and co-authored *Sacred Theatre and Walking, Autobiography and Writing*. His monograph *Spaces of Revolution: The Politics of Jean Genet's Late Theatre* will be published by Manchester University Press in 2009.

Glen McGillivray was artistic director of Australian Theatre for Young People and Theatre of Desire and now lectures in performance studies at the University of Sydney and the University of Western Sydney. He is currently researching the cultural transformation of the theatrical metaphor from the sixteenth to the eighteenth centuries.

Joseph Roach, Sterling Professor of Theater at Yale University, is the author of *The Player's Passion: Studies in the Science of Acting* (1993), *Cities of the Dead: Circum-Atlantic Performance* (1996) and *It* (2007). He is currently Principal Investigator of the World Performance Project at Yale. www.yale.edu/wpp

Bryoni Trezise lectures in theatre and performance at the School of English, Media and Performing Arts, the University of New South Wales, Sydney.